APPLIED
VECTOR
ANALYSIS

Electrical Engineering Textbook Series

Richard C. Dorf, Series Editor
University of California, Davis

Forthcoming and Published Titles

Applied Vector Analysis
Matiur Rahman and Isaac Mulolani

Continuous Signals and Systems with MATLAB
Taan ElAli and Mohammad A. Karim

Discrete Signals and Systems with MATLAB
Taan ElAli

Electromagnetics
Edward J. Rothwell and Michael J. Cloud

Optimal Control Systems
Subbaram Naidu

APPLIED VECTOR ANALYSIS

Matiur Rahman

Department of Engineering Mathematics
Dalhousie University
Halifax, Nova Scotia, Canada

Isaac Mulolani

Department of Engineering Mathematics
Dalhousie University
Halifax, Nova Scotia, Canada

CRC

CRC Press
Boca Raton   London   New York   Washington, D.C.

Library of Congress Cataloging-in-Publication Data

Rahman, M. (Matiur), 1940-
    Applied vector analysis / Matiur Rahman, Isaac Mulolani.
        p. cm.—(Electrical engineering textbook series)
    Includes bibliographical references and index.
    ISBN 0-8493-1088-1 (alk. paper)
    1. Vector analysis. I. Mulolani, Isaac. II. Title. III. Series.

QA433 .R24 2001
515'.63—dc21
        2001025920

Visit the CRC Press Web site at www.crcpress.com

Acknowledgments

The authors are thankful to the Natural Sciences and Engineering Research Council of Canada for their financial support. Thanks are also extended to Rhonda Sutherland for her work in drafting all the figures for this text. We are also grateful to CRC Press LLC for their efforts in producing a published book from our submitted manuscript.

Contents

Preface

Vector Analysis is one of the most useful branches of mathematics. It is used in practical problems arising in engineering and the applied sciences. The main purpose of this book is to illustrate the application of vector calculus to physical problems. This book is suitable for a one semester course for senior undergraduate and graduate students in science and engineering. It is also suitable for scientists and engineers working in practical fields.

The book is the outgrowth of class notes used over many years of teaching vector analysis in many universities including DalTech, Dalhousie University. The theory is explained clearly and there is an abundance of worked examples throughout. The beauty of the book is its richness of examples. There are seven chapters each of which contains ample exercises at the end. Solutions to selected exercises have been included in Appendix A. Appendix B contains a quick summary of important vector formulae.

Chapter 1 contains a brief historical background of vector calculus, which was originally introduced in the days of Aristotle and further developed in the fifteenth century. A systematic development of the evolution of vectors and basic vector algebra is described. Modern vector calculus and vector analysis are due to the contributions of Gibbs (1839–1903) and Heaviside (1850–1925). Later, Maxwell contributed heavily to the electromagnetic theory applications.

Chapter 2 introduces the algebra of vectors. This chapter contains vector addition and subtraction, scalar and vector products and vector triple products. Representation of vectors by the coordinate system and equations of lines and planes using vector algebra are demonstrated. Partial derivatives of a scalar function for several independent variables are also considered and the idea is extended to vector functions. The chapter is concluded with a number of examples of practical interest.

In Chapter 3, we consider vector functions of one variable to illustrate vector differentiation and integration along with the geometric interpretation of a position vector. Vector integration is applied to determine the length and arc length of a curve. This chapter also contains particle motion on a curve, its velocity and acceleration. Tangential and normal components of acceleration of a particle moving in space are discussed in a clear manner. The chapter ends with a considerable number of practical exercises.

The Del ∇ operator plays a very important role in practical problems arising in engineering and physical sciences. Chapter 4 is devoted to the precise definition of this important operator. With the help of this operator, a number of quantities such as gradient, divergence and curl are defined in the light of physical applications. Applied problems such as continuity of a fluid flow in incompressible and compressible fluids and also rotation of a fluid in the viscous case are very richly described with physical interpretation. We have also cited some formulae involving the Del operator with proofs.

The Laplace operator ∇^2 plays a very significant role in physical problems and is discussed very thoroughly. The chapter ends with exercises containing a number of important problems.

In Chapter 5, we have discussed the mathematical theory containing line, surface and volume integrals starting with the basics. A conservative vector field has been defined and illustrated with examples. Surface and volume integrals are also discussed with some practical examples. Triple integrals in cylindrical and spherical coordinates are discussed in this chapter. Practical situations are illustrated. Ample problems are included at the end of the chapter.

Chapter 6 contains some important and extremely useful integral theorems, such as Green's Theorem, Stokes' Theorem and the Divergence Theorem. Many examples are solved to help the readers understand the practical use of these sophisticated theorems. They are very useful for the further study of research problems. The chapter concludes with a number of exercises of academic and physical interest.

The last chapter (Chapter 7) of this book contains the applications of vector analysis. We have selected problems in fluid flow, electromagnetic theory with emphasis on Maxwell's equations. Some solution techniques are considered for Maxwell's wave and heat equations. A brief discussion on nonlinear ocean wave interaction has been presented at the end to show why vector analysis is important to the solution of physical problems. We have briefly demonstrated how to solve the nonlinear wave energy equation. The chapter ends with a number of exercises of physical interest. The last section of the chapter deals with the numerical simulation of a vector field using *Mathematica 3.0*.

We have provided a bibliography and two appendices; Appendix A contains answers to some selected exercises and Appendix B contains vector formulae at a glance.

Matiur Rahman and Isaac Mulolani

Chapter 1

Historical Background

1.1 Introduction

In our natural world there are physical quantities that are associated with only a magnitude. These can be represented by real numbers. Some examples are temperature, density, volume, moment of inertia, speed and pressure. They are known as **scalars**. However, numerous other quantities have associated with them both magnitude and direction, and thus cannot be described by a single number. Velocity, acceleration, force, and electric current are the most obvious concepts in this category. Representation of such quantities mathematically necessitates the introduction of **vectors**.

Derived from the Latin for "to carry," the word **vector** and the idea that a force is a vector quantity — in that the resultant of two forces can be found using the parallelogram law — was known to Aristotle. Galileo Galilei (1564–1642) stated the parallelogram law explicitly. While the roots of our subject lie far back in history, vector analysis as we know it today is largely a nineteenth century creation. As a result of work of Wessel in the late eighteenth century, the use of complex numbers to represent vectors in a plane was a well established result by the year 1830.

If several vectors act on a body these forces will not necessarily all lie in the same plane. This realization led to a search for a "three dimensional complex number" and an associated algebra. There were two significant developments in this regard. The first was the invention of **quaternions** by W. R. Hamilton (1805–65) in Ireland in 1843 and the second was the publication of the book *Die linearle Ausdehnungslehre (The Calculus of Extensions)* by H. G. Grassmann (1809–77) in Germany in 1844. These events were both independent and motivated by entirely different considerations even though they were almost simultaneous. Hamilton's desire was to develop mathematical tools for use in astronomy and physics, while Grassmann's pursuits were more abstract and philosophical. Another of Grassmann's concerns was to develop a theoretical algebraic structure for multidimensional geometry.

1.2   Hamilton's Quaternions

A quaternion, as Hamilton defined it, is an entity of the form

$$a + b\mathbf{i} + c\mathbf{j} + d\mathbf{k}$$

where a, b, c and d are scalars and $\mathbf{i}$, $\mathbf{j}$ and $\mathbf{k}$ are somewhat analogous to the quantity $\mathbf{i}$ in complex number theory. The quantity a is referred to as the scalar part of the quaternion and the remainder as the vector part. The three coefficients b, c and d are Cartesian coordinates of a point $\mathbf{P}$. The operation of adding two quaternions is defined in an obvious way, and the criterion for equality of two quaternions is as one would expect. The usual algebraic rules of multiplication are assumed to be valid, except that the rules for the products of $\mathbf{i}$, $\mathbf{j}$ and $\mathbf{k}$ are

$$\mathbf{i}^2 = \mathbf{j}^2 = \mathbf{k}^2 = -1 \quad \text{(hence the similarity to complex numbers)}$$
$$\mathbf{ij} = \mathbf{k}, \quad \mathbf{ji} = -\mathbf{k}$$
$$\mathbf{jk} = \mathbf{i}, \quad \mathbf{kj} = -\mathbf{i}$$
$$\mathbf{ki} = \mathbf{j}, \quad \mathbf{ik} = -\mathbf{j}$$

where the commutative law is abandoned. In these we can see the similarity with the familiar cross product results. Hence, if

$$\alpha = \alpha_0 + \alpha_1\mathbf{i} + \alpha_2\mathbf{j} + \alpha_3\mathbf{k} \quad \text{and} \quad \beta = \beta_0 + \beta_1\mathbf{i} + \beta_2\mathbf{j} + \beta_3\mathbf{k}$$

are two quaternions, then their product is

$$\begin{aligned}
\alpha\beta = {}&\alpha_0\beta_0 - (\alpha_1\beta_1 + \alpha_2\beta_2 + \alpha_3\beta_3) + (\alpha_2\beta_3 - \beta_2\alpha_3)\mathbf{i} \\
&+ (\alpha_3\beta_1 - \beta_3\alpha_1)\mathbf{j} + (\alpha_1\beta_2 - \beta_1\alpha_2)\mathbf{k} \\
&+ (\alpha_0\beta_1 + \beta_0\alpha_1)\mathbf{i} + (\alpha_0\beta_2 + \beta_0\alpha_2)\mathbf{j} \\
&+ (\alpha_0\beta_3 + \beta_0\alpha_3)\mathbf{k}.
\end{aligned} \tag{1.1}$$

Hamilton also introduced the symbol ∇, to denote the differential operator

$$\mathbf{i}\frac{\partial}{\partial x} + \mathbf{j}\frac{\partial}{\partial y} + \mathbf{k}\frac{\partial}{\partial z}$$

which, when applied to a scalar point function $\phi(x, y, z)$, produces a vector

$$\nabla\phi = \mathbf{i}\frac{\partial\phi}{\partial x} + \mathbf{j}\frac{\partial\phi}{\partial y} + \mathbf{k}\frac{\partial\phi}{\partial z}.$$

Application of this operator to a vector point function

$$\mathbf{B} = b_1\mathbf{i} + b_2\mathbf{j} + b_3\mathbf{k}$$

produced a quaternion of the form:

$$
\begin{aligned}
\nabla\mathbf{B} &= \left(\mathbf{i}\frac{\partial}{\partial x} + \mathbf{j}\frac{\partial}{\partial y} + \mathbf{k}\frac{\partial}{\partial z} \right)(b_1\mathbf{i} + b_2\mathbf{j} + b_3\mathbf{k}) \\
&= -\left(\frac{\partial b_1}{\partial x} + \frac{\partial b_2}{\partial y} + \frac{\partial b_3}{\partial z} \right) + \left(\frac{\partial b_3}{\partial y} - \frac{\partial b_2}{\partial z} \right)\mathbf{i} \\
&\quad + \left(\frac{\partial b_1}{\partial z} - \frac{\partial b_3}{\partial x} \right)\mathbf{j} + \left(\frac{\partial b_2}{\partial x} - \frac{\partial b_1}{\partial y} \right)\mathbf{k}
\end{aligned}
$$

following the multiplication rules. In our present understanding, we identify the scalar part of $\nabla\mathbf{B}$ as the negative of the divergence of $\mathbf{B}$ and the vector part as the **curl** of $\mathbf{B}$.

Hamilton's application of quaternions was in solving physical problems. For the remainder of his life he strongly encouraged their use to other physicists. P. G. Tait (1831–1901) of Scotland pressed for their use as a fundamental tool in both physics and geometry. Despite these efforts, quaternions were not extensively utilized by either physicists or applied mathematicians. Conventional Cartesian coordinates continued to be the preferred tool.

1.3   Grassmann's Calculus of Extension

The extensive quantity of Grassmann's basic concept has n components,

$$\alpha_1 e_1 + \alpha_2 e_2 + \ldots + \alpha_n e_n$$

where α_i are scalars and e_i are primary. In the case of $n = 3$, we have geometric representations as directed line segments of unit length drawn from a common origin to determine an orthogonal right-handed system of axes. In addition to the obvious sum of two such quantities, Grassmann formulated two kinds of multiplication:

1. An inner product, such that

$$e_i/e_i = 1 \quad \text{and} \quad e_i/e_j = 0, i \neq j.$$

so that if $\alpha = \alpha_1 e_1 + \alpha_2 e_2 + \ldots + \alpha_n e_n$

and $\beta = \beta_1 e_1 + \beta_2 e_2 \ldots + \beta_n e_n$ the inner product $[\alpha/\beta]$ is defined as

$$[\alpha/\beta] = \alpha_1\beta_1 + \alpha_2\beta_2 + \ldots + \alpha_n\beta_n \qquad (1.2)$$

and $[\beta/\alpha] = [\alpha/\beta]$. Note that the usual notation "/" represents the inner product and not division here. The numerical value of one of Grassmann's extensive quantities was defined as

$$\sqrt{(\alpha/\alpha)} = \sqrt{(\alpha_1^2 + \alpha_2^2 + \ldots + \alpha_n^2)}.$$

2. An outer product $[\alpha/\beta]$, defined such that

$$[e_i e_i] = 0 \quad \text{and} \quad [e_i e_j] = -[e_j e_i], i \neq j$$

so that, in the $n = 3$ case,

$$[\alpha\beta] = (\alpha_2\beta_3 - \alpha_3\beta_2)[e_2 e_3] + (\alpha_3\beta_1 - \alpha_1\beta_3)[e_3 e_1]$$
$$+ (\alpha_1\beta_2 - \alpha_2\beta_1)[e_1 e_2]. \qquad (1.3)$$

Comparison of Grassmann's inner product (1.2), in the case where $n = 3$, with Hamilton's quaternion product (1.1) reveals that the former is equivalent to the negative of the scalar part of Hamilton's quaternion product when the two quaternions have no scalar part (i.e., when they are vectors). In the three-dimensional case, if we replace $[e_2 e_3]$ by e_1, and so on, Grassmann's outer product (1.3) is equivalent to the vector part of the quaternion product of two vectors. An essential difference between the two is that in Grassmann's algebra the vector is the basic quantity, while in quaternion theory it is just one part of the quaternion.

Grassmann's extension to the n-dimensional case introduced additional products and this thinking helped lead to more advanced concepts such as tensors. Despite the originality of his work, it remained little known for many years after its publication.

1.4   The Work of Maxwell

Because Grassmann's work was in obscurity and the quaternion concept considered unsuitable for applications, other developments were required. J. C. Maxwell (1831–79) in England provided the next significant development in vector analysis while working on his theoretical analysis of electricity and magnetism. Being familiar with Hamilton's quaternion concept, he separated its vector and scalar part. Maxwell denoted the scalar part of the quaternion by $S\nabla \mathbf{B}$ and called it the **convergence** of $\mathbf{B}$ because the expression was already in fluid dynamics. Shortly afterwards, W. L. Clifford (1845–79) coined the term **divergence** and defined

$$\text{div} \quad \mathbf{B} = -S\nabla\mathbf{B} = \frac{\partial b_1}{\partial x} + \frac{\partial b_2}{\partial y} + \frac{\partial b_3}{\partial z}.$$

Maxwell denoted the vector part by $V\nabla\mathbf{B}$ and called it the **curl** or **rotation** of $\mathbf{B}$. The choice of this term was attributed to the fact that it was already in use in fluid dynamics as twice the rate of rotation of the fluid at a point. Today, the term **rot B** still appears occasionally as a synonym for both **curl B** and $\nabla \times \mathbf{B}$. Maxwell also pointed out that the operator ∇ repeated gives

$$\nabla^2 = -\left(\frac{\partial^2}{\partial x^2} + \frac{\partial^2}{\partial y^2} + \frac{\partial^2}{\partial z^2} \right)$$

which he called the Laplacian operator. He went on to note the results

1. curl grad $u = \mathbf{0}$ for any scalar function u

2. div curl B $= 0$ for any vector field $\mathbf{B}$.

1.5 Modern Vector Analysis

J. W. Gibbs (1839–1903), in the United States, and O. Heaviside (1850–1925), an electrical engineer in Britain, made the final break from quaternion theory into the beginning of modern vector analysis. Their work was independent but the end results were effectively the same, apart from differences in notation and terminology. Gibbs, a professor of mathematical physics, was motivated by the need for a simpler mathematical framework than quaternions, both for teaching and research purposes. In 1881 he produced a pamphlet, *Elements of Vector Analysis* for private circulation among his students and other interested parties. Later, in 1901, he consented to the formal presentation of his work in book form, the actual writing of this book being done by one of his pupils, E. B. Wilson.

Most texts on vector analysis adhere to the form set down by Gibbs, although he himself was somewhat reluctant to publish his ideas. Heaviside, like Maxwell, was primarily interested in electricity and magnetism. He rejected quaternions as being unsuitable mathematical tools for this area and developed his own vector analysis, which he regarded as merely shorthand.

The correspondence between Gibbs and Heaviside, on the one hand, and Tait, as a "quaternionist," on the other makes for interesting and even amusing reading today. Heaviside was particularly acerbic in his remarks about quaternions. Some of the correspondence is quoted in the introduction to the classic text of C. E. Weatherburn — first published in 1921.

It is interesting to note that Maxwell's equations for the electromagnetic field were usually written out by Maxwell himself in component form; it was Heaviside who wrote them in vector form. The notation $\mathbf{A} \times \mathbf{B}$ for the vector product and $\nabla \times \mathbf{A}$ for the curl was used by the Italian vector analysts, particularly R. Marcolongo (1862–1943) and C. Burali-Forti (1861–1931). The problem of notation in vector analysis was a subject for heated debate in the early years of the twentieth century.

1.6   Other Contributions

The two basic integral theorems — that of Stokes, and the divergence theorem — were not originally developed in the concise language of vector analysis, indeed they predated the work of Gibbs and Heaviside by many years. In a paper published in 1831 on solving the heat conduction equation, Russian, M. Ostogradsky (1801–61) converted a volume integral of the form

$$\int \int \int_V \left(\frac{\partial P}{\partial x} + \frac{\partial Q}{\partial y} + \frac{\partial R}{\partial z} \right) dx\, dy\, dz$$

into a surface integral

$$\int \int_S (P \cos \lambda + Q \cos \mu + R \cos \nu)\, dS$$

where P, Q and R are scalar functions of x, y and z and are components of a vector and λ, μ and ν are the direction cosines of the normal to the surface S enclosing the volume V. However, the theorem, which in modern notation is, of course,

$$\int \int \int_V \nabla \cdot \mathbf{V} dV = \int \int_S \mathbf{V} \cdot \mathbf{n} dS$$

is also attributed to C. F. Gauss (1777–1855) in Germany, while George Green (1793–1841) in England also made systematic use of the integral identities equivalent to the divergence theorem in 1828. Stokes' theorem was probably first stated in a letter to Stokes by Lord Kelvin (Sir William Thomson) in England in 1850. Stokes (1819–1903) used the theorem as a question in a prize examination at Cambridge in 1854. Interestingly enough, one of the candidates for the prize was J. C. Maxwell.

Chapter 2

The Algebra of Vectors

The starting point for the definition of a vector will be the intuitive one encountered in elementary physics. As mentioned in the previous chapter, there are physical quantities that have associated with them only magnitude. These can be represented by real numbers and are called **scalars**. There are many quantities, however, that have associated with them both magnitude and direction. Such quantities cannot be completely described by a single real number. The most notable concepts in this category are velocity, acceleration and force. To represent such quantities mathematically, we introduce **vectors**.

DEFINITION 2.1   *A vector is defined as a directed line segment.*

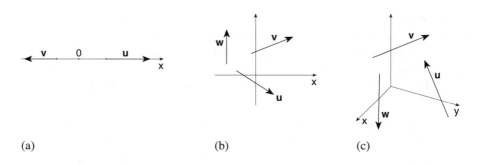

(a)   (b)   (c)

FIGURE 2.1
Vectors as directed line segments

The length of the vector will be denoted by the word **magnitude**. Any physical element that has magnitude and direction and hence can be represented as a vector, will also be designated as a vector. In Figures 2.1 (a), (b) and (c), we show two vectors **A** and **B** along a line, three vectors **A**, **B** and **C** in a plane, and three vectors **A**, **B** and **C** in space, respectively. It is customary to place an arrowhead on a vector and call this end the **tip** of the vector. The other end is called the **tail** of the vector and the direction of the vector is from tail to tip.

To distinguish vectors from scalars we shall consistently denote the former by symbols, either in upper-case or lower-case letters, set in boldface type thus, **A**. When written by hand, a vector quantity is indicated by putting an arrow above the appropriate symbol. A scalar quantity can be represented by a mark on a fixed scale. However, to represent a vector, we must use a directed line segment whose direction is the same as the direction of the vector and whose length is equal to the magnitude of the vector.

For convenience, the representative line segments shall be referred to as the vector itself. The magnitude or length of a vector **A** is called the **absolute value** of the vector and is indicated by either enclosing the symbol for the vector between ordinary absolute-value bars or simply by setting the symbol for the vector in ordinary rather than boldface type, i.e.,

$$A = |\mathbf{A}|.$$

The above represents the magnitude or absolute value of the vector **A**. A unit vector can then be defined in the following manner.

DEFINITION 2.2   *A vector* **A** *is said to be a unit vector if it has length equal to one unit; i.e.,* **A** *is a unit vector if*

$$|\mathbf{A}| = 1$$

It will follow that a **zero vector** will be one whose absolute value is zero. The direction of the zero vector is undefined and it is denoted by a zero set in boldface type. From Definition 2.1 there is no restriction on the point of application of a vector. We may thus place the tail of the vector in any spot we wish. This suggests the following definition for equality of vectors.

DEFINITION 2.3   *Two vectors whose magnitudes and directions equal are said to be equal, regardless of the points in space from which they may be drawn. If two vectors have the same length, but are oppositely directed, each is said to be the negative of the other.*

From this it can be seen that a vector quantity can be represented equally well by any of infinitely many equivalent line segments, all having the same length and direction. It is therefore customary to say that *a vector can be moved parallel to itself without change.* However, in some applications, e.g., in dealing with forces whose points of application or lines of action cannot be shifted, it is necessary to think of a vector as fixed or at least restricted in position. Such vectors are usually said to be **bound** in contrast to unrestricted vectors, which are said to be **free**.

2.1 Addition and Subtraction of Vectors

Suppose that there are two vectors, $\mathbf{A}$ and $\mathbf{B}$, which are drawn from the same point. The the **sum** of the two vectors $\mathbf{A}$ and $\mathbf{B}$ is defined by the familiar parallelogram law, i.e.,

DEFINITION 2.4   *If two vectors $\mathbf{A}$ and $\mathbf{B}$ are drawn from the same point or origin, and if the parallelogram having $\mathbf{A}$ and $\mathbf{B}$ as adjacent sides is constructed, then the sum $\mathbf{A} + \mathbf{B}$ is the vector represented by the diagonal of the parallelogram that passes through the common origin of $\mathbf{A}$ and $\mathbf{B}$ (see Figure 2.2 (a)).*

From the above definition, it is clear that,

$$\mathbf{A} + \mathbf{B} = \mathbf{B} + \mathbf{A}$$

i.e., that *vector addition is commutative*, and that,

$$\mathbf{A} + (\mathbf{B} + \mathbf{C}) = (\mathbf{A} + \mathbf{B}) + \mathbf{C}$$

i.e., that *vector addition is associative*. By the **difference** of two vectors $\mathbf{A}$ and $\mathbf{B}$, we mean the sum of the first and the negative of the second (see Figure 2.2 (b)); i.e.,

$$\mathbf{A} - \mathbf{B} = \mathbf{A} + (-\mathbf{B}).$$

To multiply a vector $\mathbf{A}$ by a scalar α we will introduce the vector $\alpha\mathbf{A} = \mathbf{A}\alpha$, whose length is equal to the product of $|\alpha|$, and the magnitude of $\mathbf{A}$, and whose direction is the same as the direction of $\mathbf{A}$ if α is positive and opposite to it if α is negative. It is not difficult to show that the following rules hold for scalar multiplication:

$$\lambda(\mathbf{A} + \mathbf{B}) = \lambda\mathbf{A} + \lambda\mathbf{B}$$
$$(\lambda + \mu)\mathbf{A} = \lambda\mathbf{A} + \mu\mathbf{A}.$$

A further consequence of the above definition is that

(a)   (b)

FIGURE 2.2
The addition and subtraction of vectors

$$\frac{\mathbf{A}}{|\mathbf{A}|} = \frac{\mathbf{A}}{A}$$

is a unit vector in the direction of $\mathbf{A}$, where $A \neq 0$.

2.2   Scalar and Vector Products

Multiplication of two vectors can be achieved in two types of products for vectors, one which yields a scalar and the other a vector.

DEFINITION 2.5   *The scalar product (dot product or inner product) of two vectors $\mathbf{A}$ and $\mathbf{B}$ is equal to the product of their absolute values and the cosine of the angle between their (positive) directions, i.e.,*

$$\mathbf{A} \cdot \mathbf{B} = |\mathbf{A}||\mathbf{B}| \cos \theta = AB \cos \theta.$$

Observe that $|\mathbf{A}| \cos \theta$ is the projection of vector $\mathbf{A}$ in the direction of $\mathbf{B}$, while $|\mathbf{B}| \cos \theta$ is the projection of the vector $\mathbf{B}$ in the direction of $\mathbf{A}$. This leads to the conclusion that *the dot product of two vectors is equal to the length of either of them multiplied by the projection of the other upon it* (see Figure 2.3). Several important cases are worth noting. First, we have seen that $\mathbf{A}/A$ is a unit vector in the direction of $\mathbf{A}$. From this, it follows that

$$\frac{\mathbf{A}}{A} \cdot \mathbf{B}$$

is the **scalar projection** of **B** in the direction of **A** and that

$$\left(\frac{\mathbf{A}}{A} \cdot \mathbf{B}\right)\frac{\mathbf{A}}{A} = \left(\frac{\mathbf{A} \cdot \mathbf{B}}{A^2}\right)\mathbf{A}$$

is the **vector projection** of **B** in the direction of **A**. If $\mathbf{B} = \mathbf{A}$, then

$$\mathbf{A} \cdot \mathbf{A} = |\mathbf{A}|^2 = A^2. \tag{2.1}$$

Of the operation rules, the associative rule is valid for the scalar product only in the wider sense, whereas the commutative and distributive rules are always valid. The scalar product is commutative,

$$\mathbf{A} \cdot \mathbf{B} = \mathbf{B} \cdot \mathbf{A}$$

associative concerning a scalar factor,

$$\lambda(\mathbf{A} \cdot \mathbf{B}) = (\lambda\mathbf{A}) \cdot \mathbf{B} = \mathbf{A} \cdot (\lambda\mathbf{B})$$

and distributive

$$(\mathbf{A} + \mathbf{B}) \cdot \mathbf{C} = \mathbf{A} \cdot \mathbf{C} + \mathbf{B} \cdot \mathbf{C}.$$

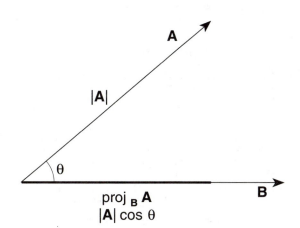

FIGURE 2.3
The geometrical interpretation of the scalar product

From the distributive property it follows that scalar multiplication of polynomial expressions is done in the same way as that of real numbers, i.e.,

$$(\mathbf{A} + \mathbf{B}) \cdot (\mathbf{C} + \mathbf{D}) = \mathbf{A} \cdot \mathbf{C} + \mathbf{B} \cdot \mathbf{C} + \mathbf{A} \cdot \mathbf{D} + \mathbf{B} \cdot \mathbf{D}.$$

These statements follow directly from the definitions. The third type of product is the **vector** or **cross product** of two vectors, indicated by putting a cross $\times$ between the vectors.

DEFINITION 2.6   *The vector product of two vectors* $\mathbf{A}$ *and* $\mathbf{B}$ *is another vector. It is denoted by* $\mathbf{A} \times \mathbf{B}$. *Its absolute value is*

$$|\mathbf{A} \times \mathbf{B}| = |\mathbf{A}||\mathbf{B}|| \sin \phi|$$

where ϕ *is the angle between the vectors* $\mathbf{A}$ *and* $\mathbf{B}$ *and its direction is perpendicular to the vectors* $\mathbf{A}$ *and* $\mathbf{B}$ *so that* $\mathbf{A}$, $\mathbf{B}$, $\mathbf{A} \times \mathbf{B}$ *form a right-hand system.*

In other words, if we measure $\mathbf{A}$, $\mathbf{B}$, $\mathbf{A} \times \mathbf{B}$ from a common starting point, the rotation at an angle smaller than π carrying the vector $\mathbf{A}$ into the direction of the vector $\mathbf{B}$ is positive if we look at it from the half-space limited by the plane of vectors $\mathbf{A}$ and $\mathbf{B}$, towards which $\mathbf{A} \times \mathbf{B}$ points (see Figure 2.4 (a)). The geometrical meaning of the absolute value of the vector product is the area of the parallelogram spanned by the two vectors.

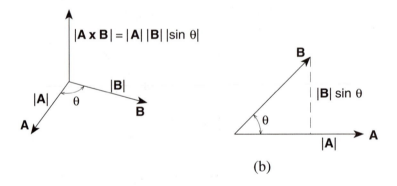

(a)   (b)

FIGURE 2.4
The geometrical interpretation of the vector product

The vector product — just like the scalar product — may be zero even if none of its vectors are zero vectors. Namely, if two vectors are parallel, the angle enclosed is 0 or π and so, on account of $\sin \phi = 0$, the vector product is 0. And, conversely, if $\mathbf{A} \times \mathbf{B} = \mathbf{0}$ and $\mathbf{A} \neq 0$, $\mathbf{B} \neq 0$, then the two vectors are parallel.

It follows then that a necessary and sufficient condition for two vectors to be parallel is that

$$\mathbf{A} \times \mathbf{B} = 0.$$

Among the operations, neither the commutative nor the associative is valid. Only the distributive rule is valid for the vector product. However, because the direction of $\mathbf{A} \times \mathbf{B}$ is determined by the right-hand rule, it is clear that interchanging $\mathbf{A}$ and $\mathbf{B}$ reverses the direction, or sign, of their product. Hence, *cross multiplication is not commutative* and we have, in fact,

$$\mathbf{A} \times \mathbf{B} = -\mathbf{B} \times \mathbf{A}.$$

Multiplication in which products obey this rule is sometimes said to be **anti-commutative**. From the foregoing it is clear that we must be careful to preserve the proper order of vectors in any expression involving vector multiplication. Concerning the multiplication of scalars, the vector product is associative:

$$\alpha(\mathbf{A} \times \mathbf{B}) = (\alpha\mathbf{A}) \times \mathbf{B}$$

or distributive:

$$(\mathbf{A} + \mathbf{B}) \times \mathbf{C} = \mathbf{A} \times \mathbf{C} + \mathbf{B} \times \mathbf{C}$$

From the distributivity it follows that polynomial expressions are to be multiplied just as real numbers:

$$(\mathbf{A} + \mathbf{B}) \times (\mathbf{C} + \mathbf{D}) = \mathbf{A} \times \mathbf{C} + \mathbf{B} \times \mathbf{C} + \mathbf{A} \times \mathbf{D} + \mathbf{B} \times \mathbf{D}.$$

The scalar factor can be factored out of any factor of the scalar product:

$$\lambda\mathbf{A} \times \mu\mathbf{B} = \lambda\mu(\mathbf{A} \times \mathbf{B})$$

and

$$\lambda(\mathbf{A} \times \mathbf{B}) = (\lambda\mathbf{A} \times \mathbf{B}) = \mathbf{A} \times (\lambda\mathbf{B}).$$

2.3 Scalar and Vector Projections

We may use the definition of the inner product to obtain the **scalar projection** of one vector onto another. Recall the projection of a line segment of a units in length to a line $a\cos\theta$, where θ is the angle between the two lines (see Figure 2.5).

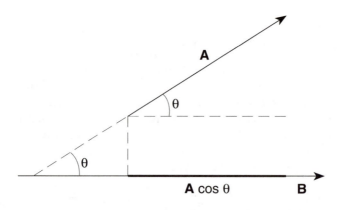

FIGURE 2.5
The scalar projection

If we wish the scalar projection of **A** onto **B**, symbolized proj$_\mathbf{B}$ **A**, we use the magnitude of **A** and the angle between **A** and **B** (Figure 2.6). This leads us to the inner product of **A** and **B**. Because

$$\mathbf{A} \cdot \mathbf{B} = |\mathbf{A}||\mathbf{B}| \cos \theta \qquad (2.2)$$

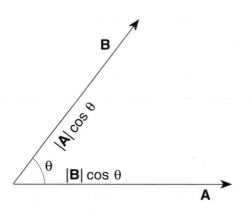

FIGURE 2.6
Geometrical interpretation of the scalar product

then

$$|\mathbf{A}| \cos \theta = \frac{\mathbf{A} \cdot \mathbf{B}}{|\mathbf{B}|}.$$

But

$$\text{proj}_\mathbf{B} \mathbf{A} = |\mathbf{A}| \cos \theta$$

thus,

$$\text{proj}_{\mathbf{B}}\mathbf{A} = \frac{\mathbf{A} \cdot \mathbf{B}}{|\mathbf{B}|}.$$

Note: If the angle θ is such that $0 \leq \theta < \frac{\pi}{2}$, the projection is positive and if $\frac{\pi}{2} < \theta < \pi$, the projection is negative. If $\theta = \frac{\pi}{2}$, the projection is zero and the vectors are perpendicular.

We may then wish to find the **vector projection** of $\mathbf{A}$ onto $\mathbf{B}$ in the direction of $\mathbf{B}$. That is, we may wish to find a vector parallel to $\mathbf{B}$, in the same direction as $\mathbf{B}$ and of a magnitude that is the absolute value of the scalar projection of $\mathbf{A}$ onto $\mathbf{B}$. We symbolize the vector projection of $\mathbf{A}$ onto $\mathbf{B}$ by $\text{proj}_{\mathbf{B}}\mathbf{A}$. To find $\text{proj}_{\mathbf{B}}\mathbf{A}$, we find $|\text{proj}_{\mathbf{B}}\mathbf{A}|$, then multiply this scalar by a unit vector in the direction of $\mathbf{B}$. Thus,

$$\begin{aligned}
\text{proj}_{\mathbf{B}}\mathbf{A} &= |\text{proj}_{\mathbf{B}}\mathbf{A}| \frac{\mathbf{B}}{|\mathbf{B}|} \\
&= \frac{|\mathbf{A} \cdot \mathbf{B}|}{|\mathbf{B}|} \frac{\mathbf{B}}{|\mathbf{B}|} \\
&= \frac{|\mathbf{A} \cdot \mathbf{B}|}{|\mathbf{B}||\mathbf{B}|} \mathbf{B} \\
&= \frac{|\mathbf{A} \cdot \mathbf{B}|}{\mathbf{B} \cdot \mathbf{B}} \mathbf{B}.
\end{aligned}$$

2.4 Cartesian Frame of Reference

2.4.1 Definition of a Position Vector R

It is often convenient to be able to refer to vector expressions in a cartesian frame of reference. To provide for this, we define $\mathbf{i}$, $\mathbf{j}$ and $\mathbf{k}$ to be vectors of unit length, directed, respectively, along the positive x, y and z axes of a right-handed rectangular coordinate system. Then $x\mathbf{i}$, $y\mathbf{j}$ and $z\mathbf{k}$ represent vectors of length $|x|$, $|y|$ and $|z|$, which run parallel to the respective axes. Thus, from the definition of vector addition it is evident that the vector joining the origin to a general point $P(x, y, z)$ (Figure 2.7) can be written

$$\mathbf{R} = x\mathbf{i} + y\mathbf{j} + z\mathbf{k}$$

that is usually known as the position vector. In the most general case, we introduce the following definition.

DEFINITION 2.7   *Let* **i**, **j**, **k** *be unit vectors in the direction of the positive axes of the rectangular three-dimensional coordinate system. Any vector,* **V**, *of three-dimensional space can be represented in the form:*

$$\mathbf{V} = v_1\mathbf{i} + v_2\mathbf{j} + v_3\mathbf{k}$$

v_1, v_2, v_3 *are called the components of the vector* **V**. *The vectors*

$$v_1\mathbf{i} = v_x, \quad v_2\mathbf{j} = v_y, \quad v_3\mathbf{k} = v_z$$

are called the projection vectors of the vector **V**.

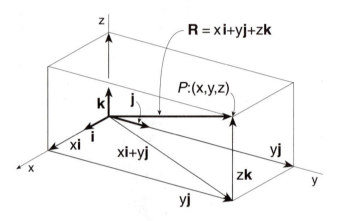

FIGURE 2.7
Vector representation using the unit vectors i, j, k

If one of the components of the vector $\mathbf{V} = v_1\mathbf{i} + v_2\mathbf{j} + v_3\mathbf{k}$ is 0 then the vector lies in the plane formed by the other two coordinate axes (the so-called two-dimensional plane).

2.5   Vector Algebra Using Coordinates

In order to solve geometric problems, it is often helpful to represent them algebraically. We now show how vectors can be represented algebraically. Suppose we denote by **AB** the vector from point A to point B. If A and B have coordinates (a_1, a_2, a_3) and (b_1, b_2, b_3) in the coordinate system shown, then the length of **AB** is

$$|\mathbf{AB}| = \sqrt{(b_1 - a_1)^2 + (b_2 - a_2)^2 + (b_3 - a_3)^2} \tag{2.3}$$

Note that if we start at the point A, proceed $b_1 - a_1$ units in the x-direction, then $b_2 - a_2$ in the y-direction, and finally $b_3 - a_3$ in the z-direction, then we arrive at B. In other words, the three numbers $b_1 - a_1$, $b_2 - a_2$, and $b_3 - a_3$ characterize both the direction and the length of the vector joining A to B. This leads to the following definition.

DEFINITION 2.8   *If the tail of a vector* **V** *is at* $A(a_1, a_2, a_3)$ *and its tip is at* $B(b_1, b_2, b_3)$, *then* **V** *shall be represented by the triple of numbers* $b_1 - a_1$, $b_2 - a_2$, $b_3 - a_3$. *In such a case we enclose the numbers in parentheses and write*

$$\mathbf{V} = (b_1 - a_1, b_2 - a_2, b_3 - a_3). \tag{2.4}$$

The equal sign in (2.4) means "is represented by." The number $b_1 - a_1$ is called the x-**component** of **V**, $b_2 - a_2$ the y-**component** of **V**, and $b_3 - a_3$ the z-**component**. Vectors in the xy-plane have only an x- and y-component:

$$\mathbf{V} = (b_1 - a_1, b_2 - a_2) \tag{2.5}$$

where (a_1, a_2) and (b_1, b_2) are the coordinates of that tail and tip of **V**. We now have an algebraic representation for vectors. Each vector has associated with it a set of components that can be found by subtracting the coordinates of its tail from the coordinates of its tip. Conversely, given a set of real numbers (a, b, c), there is one and only one vector with these numbers as components. We now examine the basic operations on vectors using the algebraic representation of vectors.

1. ## Addition of Vectors with Coordinates

The coordinates of the sum of two vectors are equal to the sum of the respective coordinates. If

$$\mathbf{A} = a_1\mathbf{i} + a_2\mathbf{j} + a_3\mathbf{k} \quad \mathbf{B} = b_1\mathbf{i} + b_2\mathbf{j} + b_3\mathbf{k}$$

then the sum of the two vectors is given by:

$$\mathbf{A} + \mathbf{B} = (a_1 + b_1)\mathbf{i} + (a_2 + b_2)\mathbf{j} + (a_3 + b_3)\mathbf{k}.$$

For example, let $\mathbf{A} = 5\mathbf{i} + \mathbf{j} - 2\mathbf{k}$ and $\mathbf{B} = 3\mathbf{i} - 4\mathbf{j} + 6\mathbf{k}$, then the sum of the two vectors is

$$\mathbf{A} + \mathbf{B} = 8\mathbf{i} - 3\mathbf{j} + 4\mathbf{k}.$$

2. Subtraction of Vectors with Coordinates

The coordinates of the difference of two vectors are equal to the difference of the respective coordinates. If $\mathbf{A} = a_1\mathbf{i} + a_2\mathbf{j} + a_3\mathbf{k}$ and $\mathbf{B} = b_1\mathbf{i} + b_2\mathbf{j} + b_3\mathbf{k}$, then the difference between the two vectors is:

$$\mathbf{A} - \mathbf{B} = (a_1 - b_1)\mathbf{i} = (a_2 - b_2)\mathbf{j} + (a_3 - b_3)\mathbf{k}.$$

Using the same two vectors from the previous example, the difference between the two is:

$$\mathbf{A} - \mathbf{B} = 2\mathbf{i} + 5\mathbf{j} - 8\mathbf{k}.$$

3. Multiplication of Vectors in Coordinate Form with Scalars

The coordinates of the product of a scalar, λ, and a vector are the λ-fold of the coordinates of the original vector. If $\mathbf{A} = a_1\mathbf{i} + a_1\mathbf{j} + a_3\mathbf{k}$, then the vector multiplied with a scalar λ is:

$$\lambda\mathbf{A} = \lambda(a_1\mathbf{i} + a_2\mathbf{j} + a_3\mathbf{k}) = \lambda a_1\mathbf{i} + \lambda a_2\mathbf{j} + \lambda a_3\mathbf{k}.$$

Let $\mathbf{A} = -7\mathbf{i} + 4\mathbf{j} + 5\mathbf{k}$, then

$$\frac{2}{3}\mathbf{A} = -\frac{14}{3}\mathbf{i} + \frac{8}{3}\mathbf{j} + \frac{10}{3}\mathbf{k}.$$

Division of vectors by scalars is defined as multiplication with the reciprocal value of the scalar, hence the coordinates of the quotient of the scalar, λ, and the vector are equal to the $\frac{1}{\lambda}$-fold of the coordinates of the original vector.

4. Coordinate Form of the Scalar Product

The scalar product of two vectors is equal to the sum of the products of the corresponding coordinates of the two vectors. If $\mathbf{A} = a_1\mathbf{i} + a_2\mathbf{j} + a_3\mathbf{k}$ and $\mathbf{B} = b_1\mathbf{i} + b_2\mathbf{j} + b_3\mathbf{k}$, then the scalar product of the two vectors is:

$$\mathbf{A} \cdot \mathbf{B} = (a_1\mathbf{i} + a_2\mathbf{j} + a_3\mathbf{k}) \cdot (b_1\mathbf{i} + b_2\mathbf{j} + b_3\mathbf{k}) = a_1b_1 + a_2b_2 + a_3b_3.$$

Recall that two vectors will be equal if and only if their respective components are equal. It follows that any vector equation implies three scalar equations. Then, because the dot product of perpendicular vectors (unit vectors in this case) is zero, it follows that

$$\mathbf{i} \cdot \mathbf{j} \;=\; \mathbf{j} \cdot \mathbf{k} \;=\; \mathbf{k} \cdot \mathbf{i} \;=\; 0.$$

Moreover, applying (2.1) to unit vectors, $\mathbf{i}$, $\mathbf{j}$, $\mathbf{k}$, we have

$$\mathbf{i} \cdot \mathbf{i} \;=\; \mathbf{j} \cdot \mathbf{j} \;=\; \mathbf{k} \cdot \mathbf{k} \;=\; 1.$$

As an example, let $\mathbf{A} = 5\mathbf{i} + \mathbf{j} - 2\mathbf{k}$ and $\mathbf{B} = 3\mathbf{i} - 4\mathbf{j} + 6\mathbf{k}$. The scalar product of the two vectors is:

$$\mathbf{A} \cdot \mathbf{B} = 5 \cdot 3 + 1 \cdot (-4) + (-2) \cdot 6 = 15 - 4 - 12 = -1.$$

In the particular case that $\mathbf{A} = \mathbf{B}$, we have

$$\mathbf{A} \cdot \mathbf{A} = |\mathbf{A}|^2 = A^2 = a_1^2 + a_2^2 + a_3^2$$

or

$$|\mathbf{A}| = A = \sqrt{a_1^2 + a_2^2 + a_3^2}.$$

However, if the dot product is written as $\mathbf{A} \cdot \mathbf{B} = |\mathbf{A}||\mathbf{B}| \cos\theta$ then solved for $\cos\theta$, using the above expressions, we obtain the useful formula,

$$\cos\theta = \frac{a_1 b_1 + a_2 b_2 + a_3 b_3}{(\sqrt{a_1^2 + a_2^2 + a_3^2})(\sqrt{b_1^2 + b_2^2 + b_3^2})}$$

which is the familiar analytic geometry result.

5. Coordinate Form of the Vector Product

For the cross product of unit vectors, $\mathbf{i}$, $\mathbf{j}$, $\mathbf{k}$ we find at one that

$$
\begin{aligned}
\mathbf{i} \times \mathbf{i} = \mathbf{j} \times \mathbf{j} \;&=\; \mathbf{k} \times \mathbf{k} \;=\; \mathbf{0} \\
\mathbf{i} \times \mathbf{j} = -\mathbf{j} \times \mathbf{i} \;&=\; \mathbf{k} \\
\mathbf{j} \times \mathbf{k} = -\mathbf{k} \times \mathbf{j} \;&=\; \mathbf{i} \\
\mathbf{k} \times \mathbf{i} = -\mathbf{i} \times \mathbf{k} \;&=\; \mathbf{j}.
\end{aligned}
$$

Using these products along with the distributive rule, we get the vector product

$$\mathbf{A} \times \mathbf{B} = (a_1\mathbf{i} + a_2\mathbf{j} + a_3\mathbf{k}) \times (b_1\mathbf{i} + b_2\mathbf{j} + b_3\mathbf{k})$$
$$= (a_2b_3 - a_3b_2)\mathbf{i} + (a_3b_1 - a_1b_3)\mathbf{j}$$
$$+ (a_1b_2 - a_2b_1)\mathbf{k}. \tag{2.6}$$

This expression may be written in compact form as a determinant, which is easier to remember:

$$\mathbf{A} \times \mathbf{B} = \begin{vmatrix} \mathbf{i} & \mathbf{j} & \mathbf{k} \\ a_1 & a_2 & a_3 \\ b_1 & b_2 & b_3 \end{vmatrix}.$$

Suppose we let $\mathbf{V}_1 = 5\mathbf{i} + \mathbf{j} - 2\mathbf{k}$ and $\mathbf{V}_2 = 3\mathbf{i} - 4\mathbf{j} + 6\mathbf{k}$. The vector product of these vectors is as follows:

$$\mathbf{V}_1 \times \mathbf{V}_2 = \begin{vmatrix} \mathbf{i} & \mathbf{j} & \mathbf{k} \\ 5 & 1 & -2 \\ 3 & -4 & 6 \end{vmatrix} = (6 - 8)\mathbf{i} - (30 + 6)\mathbf{j} + (-20 - 3)\mathbf{k}$$
$$= -2\mathbf{i} - 36\mathbf{j} - 23\mathbf{k}.$$

Note that the anti-commutative character of the vector multiplication corresponds to the fact that interchanging two rows of a determinant changes the sign of the determinant.

2.6   Mixed Product in Coordinate Form

Products involving three rather than two vectors offer the following possibilities:

$$(\mathbf{A} \cdot \mathbf{B})\mathbf{C}, \quad \mathbf{A} \cdot (\mathbf{B} \times \mathbf{C}), \quad \mathbf{A} \times (\mathbf{B} \times \mathbf{C}).$$

1. $(\mathbf{A} \cdot \mathbf{B})\mathbf{C}$.

   This first one can be defined as a vector. Because $\mathbf{A} \cdot \mathbf{B}$ is just a scalar and thus $(\mathbf{A} \cdot \mathbf{B})\mathbf{C}$ is simply a vector whose length is $|\mathbf{A} \cdot \mathbf{B}|$ times the length of $\mathbf{C}$ and whose direction is the same as that of $\mathbf{C}$ or opposite to it, according to whether $\mathbf{A} \cdot \mathbf{B}$ is positive or negative.

2. $\mathbf{A} \cdot (\mathbf{B} \times \mathbf{C})$.

This product is known as a **scalar triple product** and we observe that the parentheses enclosing the vector $\mathbf{B} \times \mathbf{C}$ are superfluous. There is, in fact, only one alternative interpretation, namely $(\mathbf{A} \cdot \mathbf{B}) \times \mathbf{C}$, and this is meaningless because both factors in a cross product must be vectors, whereas $\mathbf{A} \cdot \mathbf{B}$ is a scalar. Therefore, no meaning except the one intended can be attached to the expression $\mathbf{A} \cdot \mathbf{B} \times \mathbf{C}$ and hence it is customary to omit the parentheses. If the vectors

$$\mathbf{A} = a_1\mathbf{i} + a_2\mathbf{j} + a_3\mathbf{k}$$
$$\mathbf{B} = b_1\mathbf{i} + b_2\mathbf{j} + b_3\mathbf{k}$$
$$\mathbf{C} = c_1\mathbf{i} + c_2\mathbf{j} + c_3\mathbf{k}$$

then the mixed product of the three vectors is

$$\mathbf{A} \cdot \mathbf{B} \times \mathbf{C} = (a_1\mathbf{i} + a_2\mathbf{j} + a_3\mathbf{k}) \cdot \begin{vmatrix} \mathbf{i} & \mathbf{j} & \mathbf{k} \\ b_1 & b_2 & b_3 \\ c_1 & c_2 & c_3 \end{vmatrix}.$$

From this, it can easily be shown that:

$$\mathbf{A} \cdot \mathbf{B} \times \mathbf{C} = \begin{vmatrix} a_1 & a_2 & a_3 \\ b_1 & b_2 & b_3 \\ c_1 & c_2 & c_3 \end{vmatrix}.$$

From (2.8), and the fact that interchanging any two rows in a determinant changes the sign of the determinant, it follows easily that $\mathbf{A} \cdot \mathbf{B} \times \mathbf{C} = \mathbf{A} \times \mathbf{B} \cdot \mathbf{C}$. In other words, in any scalar product, the dot and cross can be interchanged without altering the value of the product. For this reason it is customary to omit these symbols and write a scalar triple product simply as $[\mathbf{ABC}]$. Again, either from (2.8) or directly from the anti-commutative character of vector products, it is easy to see that $\mathbf{A} \cdot \mathbf{B} \times \mathbf{C} = -\mathbf{A} \cdot \mathbf{C} \times \mathbf{B}$ and more generally

$$[\mathbf{ABC}] = [\mathbf{BCA}] = [\mathbf{CAB}] = -[\mathbf{ACB}] = -[\mathbf{BAC}] = -[\mathbf{CBA}]. \tag{2.9}$$

Thus (2.9) asserts that any cyclic permutation of factors in a scalar triple product leaves the value of the product unchanged, whereas any permutation that reverses the original cyclic order changes the sign of the product. Geometrically, the absolute value of the triple product $\mathbf{A} \cdot \mathbf{B} \times \mathbf{C}$ is the volume of the parallelepiped having vectors $\mathbf{A}$, $\mathbf{B}$ and $\mathbf{C}$ as concurrent edges. If we regard the parallelogram having $\mathbf{B}$ and

C as adjacent sides as the base of this figure, then $\mathbf{B} \times \mathbf{C}$ is a vector perpendicular to the base and whose magnitude is the projection of $\mathbf{A}$ on $\mathbf{B} \times \mathbf{C}$ (see Figure 2.8).

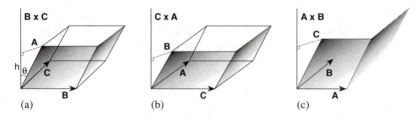

(a)   (b)   (c)

FIGURE 2.8
Geometrical interpretation of the scalar triple product

Hence, $\mathbf{A} \cdot \mathbf{B} \times \mathbf{C}$, whose value is just the magnitude of $\mathbf{B} \times \mathbf{C}$ multiplied by the projection of $\mathbf{A}$ on $\mathbf{B} \times \mathbf{C}$, is numerically equal to the volume of the parallelepiped. If the vectors, $\mathbf{A}$, $\mathbf{B}$ and $\mathbf{C}$ all lie in the same plane or are parallel to the same plane, they necessarily form a parallelepiped of zero volume and conversely. Hence, $[\mathbf{ABC}] = 0$ is a necessary and sufficient condition that the three vectors $\mathbf{A}$, $\mathbf{B}$ and $\mathbf{C}$ should be parallel to one and the same plane, i.e., coplanar. In particular, if two factors of a scalar triple product have the same direction, the product is zero. The mixed product of the vectors $\mathbf{A} = 2\mathbf{i} - \mathbf{j} + 4\mathbf{k}$, $\mathbf{B} = -\mathbf{i} + 3\mathbf{j} + \mathbf{k}$, $\mathbf{C} = -5\mathbf{i} + 2\mathbf{j} - \mathbf{k}$ is as follows:

$$[\mathbf{ABC}] = \begin{vmatrix} 2 & -1 & 4 \\ -1 & 3 & 1 \\ -5 & 2 & -1 \end{vmatrix} = 2(-3 - 2) + 1(1 + 5) + 4(-2 + 15) = 48.$$

3. $\mathbf{A} \times (\mathbf{B} \times \mathbf{C})$.

An indicated product of this form is meaningful because, by definition, $\mathbf{B} \times \mathbf{C}$ is a vector and hence is an appropriate factor in a vector product with $\mathbf{A}$. To obtain a simpler, expanded form for a product such as $\mathbf{A} \times (\mathbf{B} \times \mathbf{C})$, we first recall the expansion for $\mathbf{B} \times \mathbf{C}$:

$$\mathbf{B} \times \mathbf{C} = \begin{vmatrix} \mathbf{i} & \mathbf{j} & \mathbf{k} \\ b_1 & b_2 & b_3 \\ c_1 & c_2 & c_3 \end{vmatrix}$$
$$= (b_2 c_3 - b_3 c_2)\mathbf{i} + (b_3 c_1 - b_1 c_3)\mathbf{j} + (b_1 c_2 - b_2 c_1)\mathbf{k}$$

then use (2.7) to express the vector product of $\mathbf{A}$ and $\mathbf{B} \times \mathbf{C}$:

$$\mathbf{A} \times (\mathbf{B} \times \mathbf{C}) = \begin{vmatrix} \mathbf{i} & \mathbf{j} & \mathbf{k} \\ a_1 & a_2 & a_3 \\ (b_2c_3 - b_3c_2) & (b_3c_1 - b_1c_3) & (b_1c_2 - b_2c_1) \end{vmatrix}.$$

If we expand this determinant in terms of the (vector) elements of the first row and group positive and negative terms in each cofactor, we obtain

$$
\begin{aligned}
RHS &= \mathbf{i}[a_2(b_1c_2 - b_2c_1) - a_3(b_3c_1 - b_1c_3)] \\
&\quad + \mathbf{j}[a_3(b_2c_3 - b_3c_2) - a_1(b_1c_2 - b_2c_1)] \\
&\quad + \mathbf{k}[a_1(b_3c_1 - b_1c_3) - a_2(a_2c_3 - a_3c_2)] \\
&= \mathbf{i}[b_1(a_2c_2 + a_3c_3) - c_1(a_2b_2 + a_3b_3)] \\
&\quad + \mathbf{j}[b_2(a_3c_3 + a_1c_1) - c_2(a_3b_3 + a_1b_1)] \\
&\quad + \mathbf{k}[b_3(a_1c_1 + a_2c_2) - c_3(a_1b_1 + a_2b_2)].
\end{aligned}
$$

If we now add and subtract the products $a_1b_1c_1$, $a_2b_2c_2$, and $a_3b_3c_3$ in the respective bracketed expressions and note the dot products we have thus formed, this becomes

$$
\begin{aligned}
RHS &= \mathbf{i}[b_1(\mathbf{A} \cdot \mathbf{C}) - c_1(\mathbf{A} \cdot \mathbf{B})] \\
&\quad + \mathbf{j}[b_2(\mathbf{A} \cdot \mathbf{C}) - c_2(\mathbf{A} \cdot \mathbf{B})] \\
&\quad + \mathbf{k}[b_3(\mathbf{A} \cdot \mathbf{C}) - c_3(\mathbf{A} \cdot \mathbf{B})] \\
&= (\mathbf{A} \cdot \mathbf{C})(b_1\mathbf{i} + b_2\mathbf{j} + b_3\mathbf{k}) \\
&\quad - (\mathbf{A} \cdot \mathbf{B})(c_1\mathbf{i} + c_2\mathbf{j} + c_3\mathbf{k}) \\
&= (\mathbf{A} \cdot \mathbf{C})\mathbf{B} - (\mathbf{A} \cdot \mathbf{B})\mathbf{C}.
\end{aligned}
$$

Finally, by regrouping and factoring, this becomes the required formula:

$$\mathbf{A} \times (\mathbf{B} \times \mathbf{C}) = (\mathbf{A} \cdot \mathbf{C})\mathbf{B} - (\mathbf{A} \cdot \mathbf{B})\mathbf{C}. \tag{2.10}$$

Straightforward application of (2.10) leads to

$$(\mathbf{A} \times \mathbf{B}) \times \mathbf{C} = -\mathbf{C} \times (\mathbf{A} \times \mathbf{B}) = -(\mathbf{C} \cdot \mathbf{B})\mathbf{A} + (\mathbf{C} \cdot \mathbf{A})\mathbf{B}$$

that is *not* equal to $\mathbf{A} \times (\mathbf{B} \times \mathbf{C})$. Hence, vector multiplication is not associative as mentioned before.

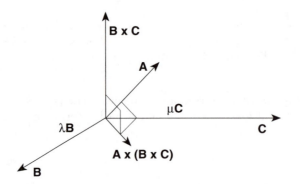

FIGURE 2.9
Geometric interpretation of the vector triple product

With a knowledge of scalar and vector triple products, products involving more than three vectors can be expanded without difficulty. For example,

$$(\mathbf{A} \times \mathbf{B}) \cdot (\mathbf{C} \times \mathbf{D})$$

can be regarded as a scalar triple product of the vectors, $\mathbf{A}$, $\mathbf{B}$ and $\mathbf{C} \times \mathbf{D}$. This allows us to write

$$\begin{aligned}
(\mathbf{A} \times \mathbf{B}) \cdot (\mathbf{C} \times \mathbf{D}) &= \mathbf{A} \cdot [\mathbf{B} \times (\mathbf{C} \times \mathbf{D})] \\
&= \mathbf{A} \cdot [(\mathbf{B} \cdot \mathbf{D})\mathbf{C} - (\mathbf{B} \cdot \mathbf{C})\mathbf{D}] \\
&= (\mathbf{A} \cdot \mathbf{C})(\mathbf{B} \cdot \mathbf{D}) - (\mathbf{A} \cdot \mathbf{D})(\mathbf{B} \cdot \mathbf{C}).
\end{aligned}$$

This result is referred to as **Lagrange's** identity. Similarly, $(\mathbf{A} \times \mathbf{B}) \times (\mathbf{C} \times \mathbf{D})$ can be thought of as the vector triple product of $\mathbf{A} \times \mathbf{B}$, $\mathbf{C}$, and $\mathbf{D}$. Then applying (2.10), we find

$$\begin{aligned}
(\mathbf{A} \times \mathbf{B}) \times (\mathbf{C} \times \mathbf{D}) &= (\mathbf{A} \times \mathbf{B} \cdot \mathbf{D})\mathbf{C} - (\mathbf{A} \times \mathbf{B} \cdot \mathbf{C})\mathbf{D} \\
&= [\mathbf{ABD}]\mathbf{C} - [\mathbf{ABC}]\mathbf{D}
\end{aligned}$$

which is a vector in the plane of $\mathbf{C}$ and $\mathbf{D}$. From another point of view, we have,

$$\begin{aligned}
(\mathbf{A} \times \mathbf{B}) \times (\mathbf{C} \times \mathbf{D}) &= -(\mathbf{C} \times \mathbf{D}) \times (\mathbf{A} \times \mathbf{B}) \\
&= -[(\mathbf{C} \times \mathbf{D} \cdot \mathbf{B})\mathbf{A} - (\mathbf{C} \times \mathbf{D} \cdot \mathbf{A})\mathbf{B}] \\
&= -(\mathbf{C} \times \mathbf{D} \cdot \mathbf{B})\mathbf{A} + (\mathbf{C} \times \mathbf{D} \cdot \mathbf{A})\mathbf{B} \\
&= -[\mathbf{CDB}]\mathbf{A} + [\mathbf{CDA}]\mathbf{B}
\end{aligned}$$

which is a vector in the plane of **A** and **B**. These two results together show that $(\mathbf{A} \times \mathbf{B}) \times (\mathbf{C} \times \mathbf{D})$ is directed along the line of intersection of the plane of **A** and **B** and the plane of **C** and **D**, provided these planes are distinct.

Example 2.1
If **A**, **B** and **C** are any three vectors, prove that

$$\mathbf{A} \times (\mathbf{B} \times \mathbf{C}) + \mathbf{B} \times (\mathbf{C} \times \mathbf{A}) + \mathbf{C} \times (\mathbf{A} \times \mathbf{B}) = \mathbf{0}.$$

▯

SOLUTION Using Lagrange's identity, the left-hand side of the above can be expressed as

$$
\begin{aligned}
LHS &= \mathbf{A} \times (\mathbf{B} \times \mathbf{C}) + \mathbf{B} \times (\mathbf{C} \times \mathbf{A}) + \mathbf{C} \times (\mathbf{A} \times \mathbf{B}) \\
&= [(\mathbf{A} \cdot \mathbf{C})\mathbf{B} - (\mathbf{A} \cdot \mathbf{B})\mathbf{C}] + [(\mathbf{B} \cdot \mathbf{A})\mathbf{C} - (\mathbf{B} \cdot \mathbf{C})\mathbf{A}] \\
&\quad + [(\mathbf{C} \cdot \mathbf{B})\mathbf{A} - (\mathbf{C} \cdot \mathbf{A})\mathbf{B}] \\
&= \mathbf{0} \\
&= RHS.
\end{aligned}
$$

Hence the required proof. ∎

Example 2.2
Prove that

$$(\mathbf{A} \times \mathbf{B}) \cdot (\mathbf{C} \times \mathbf{D}) + (\mathbf{B} \times \mathbf{C}) \cdot (\mathbf{A} \times \mathbf{D}) + (\mathbf{C} \times \mathbf{A}) \cdot (\mathbf{B} \times \mathbf{D}) = 0.$$

▯

SOLUTION Again we use Lagrange's identity as follows,

$$
\begin{aligned}
LHS &= (\mathbf{A} \cdot \mathbf{C})(\mathbf{B} \cdot \mathbf{D}) - (\mathbf{A} \cdot \mathbf{D})(\mathbf{B} \cdot \mathbf{C}) \\
&\quad + (\mathbf{B} \cdot \mathbf{A})(\mathbf{C} \cdot \mathbf{D}) - (\mathbf{B} \cdot \mathbf{D})(\mathbf{C} \cdot \mathbf{A}) \\
&\quad + (\mathbf{C} \cdot \mathbf{B})(\mathbf{A} \cdot \mathbf{D}) - (\mathbf{C} \cdot \mathbf{D})(\mathbf{A} \cdot \mathbf{B}) \\
&= 0 \\
&= RHS.
\end{aligned}
$$

Hence the required proof. ∎

Example 2.3
Prove that

$$(\mathbf{A} \times \mathbf{B}) \cdot (\mathbf{B} \times \mathbf{C}) \times (\mathbf{C} \times \mathbf{A}) = [\mathbf{ABC}]^2.$$

⬚

SOLUTION   Using the Lagrange identity, we can write the left-hand side as

$$
\begin{aligned}
LHS &= (\mathbf{A} \times \mathbf{B}) \cdot (\mathbf{B} \times \mathbf{C}) \times (\mathbf{C} \times \mathbf{A}) \\
&= (\mathbf{A} \times \mathbf{B}) \cdot [(\mathbf{B} \times \mathbf{C} \cdot \mathbf{A})\mathbf{C} - (\mathbf{B} \times \mathbf{C} \cdot \mathbf{C})\mathbf{A}] \\
&= (\mathbf{A} \times \mathbf{B}) \cdot [(\mathbf{B} \times \mathbf{C} \cdot \mathbf{A})\mathbf{C} - 0] \\
&= [\mathbf{A} \times \mathbf{B} \cdot \mathbf{C}][\mathbf{B} \times \mathbf{C} \cdot \mathbf{A}] \\
&= [\mathbf{ABC}][\mathbf{ABC}] = [\mathbf{ABC}]^2 \\
&= RHS.
\end{aligned}
$$

Hence the required proof.   ∎

Example 2.4
Verify that if

$$
\begin{aligned}
\mathbf{A} &= \mathbf{i}a_1 + \mathbf{j}a_2 + \mathbf{k}a_3 \\
\mathbf{B} &= \mathbf{i}b_1 + \mathbf{j}b_2 + \mathbf{k}b_3 \\
\mathbf{C} &= \mathbf{i}c_1 + \mathbf{j}c_2 + \mathbf{k}c_3 \\
\text{and } \mathbf{D} &= \mathbf{i}d_1 + \mathbf{j}d_2 + \mathbf{k}d_3,
\end{aligned}
$$

then the system of equations

$$
\begin{aligned}
a_1 x + b_1 y + c_1 z &= d_1 \\
a_2 x + b_2 y + c_2 z &= d_2 \\
a_3 x + b_3 y + c_3 z &= d_3,
\end{aligned}
$$

is equivalent to the single vector equation

$$x\mathbf{A} + y\mathbf{B} + z\mathbf{C} = \mathbf{D}.$$

Assuming that $[\mathbf{ABC}] \neq 0$, solve this vector equation for x, y and z. Show that the result is equivalent to that obtained from the algebraic form of the system by using Cramer's rule.   ⬚

SOLUTION   We can easily establish the linear system if we take that dot product of $x\mathbf{A} + y\mathbf{B} + z\mathbf{C} = \mathbf{D}$ with the unit vectors $\mathbf{i}$, $\mathbf{j}$ and $\mathbf{k}$.

$$\mathbf{i} \cdot (x\mathbf{A} + y\mathbf{B} + z\mathbf{C}) = \mathbf{i} \cdot \mathbf{D}$$
$$a_1 x + b_1 y + c_1 z = d_1.$$

Similarly, we can establish the other two equations. To find the solution x, y, and z of the vector equation, we proceed as follows. First, we cross multiply by $\mathbf{C}$ to obtain

$$x\mathbf{A} \times \mathbf{C} + y\mathbf{B} \times \mathbf{C} + z\mathbf{C} \times \mathbf{C} = \mathbf{D} \times \mathbf{C}$$

Because we know $\mathbf{C} \times \mathbf{C} = 0$, $x\mathbf{A} \times \mathbf{C} + y\mathbf{B} \times \mathbf{C} = \mathbf{D} \times \mathbf{C}$. Now by scalar multiplication of the resulting equation by the vector $\mathbf{B}$, we get

$$x\mathbf{A} \times \mathbf{C} \cdot \mathbf{B} + y\mathbf{B} \times \mathbf{C} \cdot \mathbf{B} = \mathbf{D} \times \mathbf{C} \cdot \mathbf{B}$$
$$x\mathbf{A} \times \mathbf{C} \cdot \mathbf{B} = \mathbf{D} \times \mathbf{C} \cdot \mathbf{B}$$

because we know that $\mathbf{B} \times \mathbf{C} \cdot \mathbf{B} = 0$. Therefore, we get

$$x\mathbf{A} \times \mathbf{C} \cdot \mathbf{B} = \mathbf{D} \times \mathbf{C} \cdot \mathbf{B}$$
$$x\mathbf{A} \cdot \mathbf{B} \times \mathbf{C} = \mathbf{D} \cdot \mathbf{B} \times \mathbf{C} \quad \text{using cyclic permutation property}$$
$$x[\mathbf{ABC}] = [\mathbf{DBC}]$$
$$x = \frac{[\mathbf{DBC}]}{[\mathbf{ABC}]}.$$

Similarly, we can obtain the values for y and z as

$$y = \frac{[\mathbf{ADC}]}{[\mathbf{ABC}]}, \quad z = \frac{[\mathbf{ABD}]}{[\mathbf{ABC}]}.$$

Solving the 3x3 system by Cramer's rule, we get

$$x = \frac{\begin{vmatrix} d_1 & b_1 & c_1 \\ d_2 & b_2 & c_2 \\ d_3 & b_3 & c_3 \end{vmatrix}}{\begin{vmatrix} a_1 & b_1 & c_1 \\ a_2 & b_2 & c_2 \\ a_3 & b_3 & c_3 \end{vmatrix}}.$$

From the vector method, we have

$$x = \frac{\begin{vmatrix} d_1 & d_2 & d_3 \\ b_1 & b_2 & b_3 \\ c_1 & c_2 & c_3 \end{vmatrix}}{\begin{vmatrix} a_1 & a_2 & a_3 \\ b_1 & b_2 & b_3 \\ c_1 & c_2 & c_3 \end{vmatrix}},$$

using the property of determinants in changing the columns to rows. The value of the determinant will not be changed if we transpose the rows and columns. ∎

Example 2.5

Using vector methods, determine the law of cosines.

SOLUTION To obtain the law of cosines, let us assume that the sides of the given triangle are $\mathbf{A}$, $\mathbf{B}$ and $\mathbf{C}$ such that $\mathbf{C} = \mathbf{A} - \mathbf{B}$. Then by taking the dot product we have the following:

$$\mathbf{C} \cdot \mathbf{C} = (\mathbf{A} - \mathbf{B}) \cdot (\mathbf{A} - \mathbf{B})$$
$$= \mathbf{A} \cdot \mathbf{A} - 2\mathbf{A} \cdot \mathbf{B} + \mathbf{B} \cdot \mathbf{B}$$
$$= |A|^2 - 2|A||B|\cos\theta + |B||B|, \quad \text{which yields}$$
$$C^2 = A^2 - 2AB\cos\theta + B^2.$$

Here θ is the angle between $\mathbf{A}$ and $\mathbf{B}$. This is the required law of cosines, which the reader is no doubt familiar with. ∎

Example 2.6

Find the unit vector which is perpendicular to both $\mathbf{A} = \mathbf{i} - 2\mathbf{j} + \mathbf{k}$ and $\mathbf{B} = 3\mathbf{i} + \mathbf{j} - 2\mathbf{k}$.

SOLUTION We know that the cross product of two vectors is a vector which is perpendicular to both of the original vectors. This means that we can write

$$\mathbf{A} \times \mathbf{B} = \begin{vmatrix} \mathbf{i} & \mathbf{j} & \mathbf{k} \\ 1 & -2 & 1 \\ 3 & 1 & -2 \end{vmatrix}$$
$$= 3\mathbf{i} + 5\mathbf{j} + 7\mathbf{k}$$
$$= \mathbf{C}$$

which is a vector perpendicular to both **A** and **B**. Then the unit vector of **C** is given by

$$\frac{\mathbf{C}}{|\mathbf{C}|} = \frac{3\mathbf{i} + 5\mathbf{j} + 7\mathbf{k}}{\sqrt{9 + 25 + 49}}$$
$$= \frac{3\mathbf{i} + 5\mathbf{j} + 7\mathbf{k}}{\sqrt{83}}$$
$$= \frac{3}{\sqrt{83}}\mathbf{i} + \frac{5}{\sqrt{83}}\mathbf{j} + \frac{7}{\sqrt{83}}\mathbf{k}.$$

∎

Example 2.7
Find the unit vector that is parallel to the plane of **A** = **i** + **j** − 2**k** and **B** = 3**i** − 2**j** + **k** and perpendicular to **C** = 2**i** + 2**j** − **k**. ▯

SOLUTION   We know that the cross product of the vectors **A** and **B** is a vector perpendicular to both **A** and **B**, i.e., **A** × **B**. We need to determine a vector that is perpendicular to both **A** × **B** and **C** such that the resulting vector will be parallel to **A** and **B** by the vector cross product. That is to say,

$$
\begin{aligned}
(\mathbf{A} \times \mathbf{B}) \times \mathbf{C} &= -\mathbf{C} \times (\mathbf{A} \times \mathbf{B}) \\
&= -(\mathbf{C} \cdot \mathbf{B})\mathbf{A} + (\mathbf{C} \cdot \mathbf{A})\mathbf{B} \\
&= -(6 - 4 - 1)(\mathbf{i} + \mathbf{j} - 2\mathbf{k}) + (2 + 2 + 2)(3\mathbf{i} - 2\mathbf{j} + \mathbf{k}) \\
&= -(\mathbf{i} + \mathbf{j} - 2\mathbf{k}) + 6(3\mathbf{i} - 2\mathbf{j} + \mathbf{k}) \\
&= 17\mathbf{i} - 13\mathbf{j} + 8\mathbf{k}.
\end{aligned}
$$

Therefore, the required unit vector is given by

$$\frac{17\mathbf{i} - 13\mathbf{j} + 8\mathbf{k}}{\sqrt{17^2 + 13^2 + 8^2}} = \frac{17\mathbf{i} - 13\mathbf{j} + 8\mathbf{k}}{\sqrt{522}}.$$

∎

2.7   Vector Representation Using Coordinates

1. Absolute Value of a Vector

The absolute value of a vector — according to the definition — is the square root of the sum of the squares of the coordinates

$$|\mathbf{A}| = \sqrt{|\mathbf{A}|^2} = \sqrt{\mathbf{A} \cdot \mathbf{A}}$$
$$= \sqrt{(a_1\mathbf{i} + a_2\mathbf{j} + a_3\mathbf{k}) \cdot (a_1\mathbf{i} + a_2\mathbf{j} + a_3\mathbf{k})}$$
$$= \sqrt{a_1^2 + a_2^2 + a_3^2}.$$

The absolute value of the vector $\mathbf{A} = 4\mathbf{i} + 2\mathbf{j} - 3\mathbf{k}$ is

$$|\mathbf{A}| = \sqrt{16 + 4 + 9} = \sqrt{29}.$$

If $\mathbf{e}$ is a unit vector, i.e., $|\mathbf{e}| = 1$ then

$$\sqrt{e_1^2 + e_2^2 + e_3^2} = 1$$

and so $e_1^2 + e_2^2 + e_3^2 = 1$. This means that the sum of the squares of the coordinates of a unit vector is equal to one. The vector $\mathbf{B} = \frac{1}{2}\mathbf{i} - \frac{5}{6}\mathbf{j} + \frac{\sqrt{2}}{6}\mathbf{k}$ is a unit vector because

$$e_1^2 + e_2^2 + e_3^2 = \frac{9}{36} + \frac{25}{36} + \frac{2}{36} = 1.$$

2. **Unit Vectors Represented by Coordinates**

A unit vector pointing in the direction of an arbitrary vector $\mathbf{B} = 5\mathbf{i} - 3\mathbf{j} + \mathbf{k}$ is

$$\mathbf{e_B} = \frac{5\mathbf{i} - 3\mathbf{j} + \mathbf{k}}{\sqrt{25 + 9 + 1}} = \frac{5}{\sqrt{35}}\mathbf{i} - \frac{3}{\sqrt{35}}\mathbf{j} + \frac{1}{\sqrt{35}}\mathbf{k} = \frac{\mathbf{B}}{|\mathbf{B}|}.$$

3. **Direction Cosines of a Vector**

The cosines of the angles enclosed by a vector and the coordinate axes (respectively, the unit vectors $\mathbf{i}$, $\mathbf{j}$, $\mathbf{k}$) are called the direction cosines of the vector (see Figure 2.10).

Denoting the angles enclosed by the vector $\mathbf{A} = a_1\mathbf{i} + a_2\mathbf{j} + a_3\mathbf{k}$ and the unit vectors, $\mathbf{i}$, $\mathbf{j}$, $\mathbf{k}$ by α, β, γ respectively, we can easily find the direction cosines of the line vector by using the dot products as follows

$$\mathbf{i} \cdot \mathbf{A} = |\mathbf{i}||\mathbf{A}| \cos \alpha$$
$$\text{or} \quad \cos \alpha = \frac{a_1}{|\mathbf{A}|} = \frac{a_1}{\sqrt{a_1^2 + a_2^2 + a_3^2}}.$$

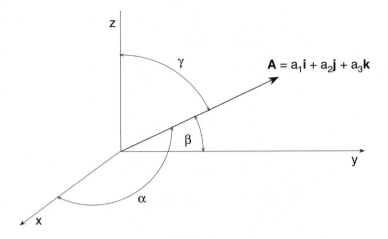

FIGURE 2.10
Direction cosines of a vector

Similarly, we can find that

$$\cos \beta = \frac{a_2}{|\mathbf{A}|} = \frac{a_2}{\sqrt{a_1^2 + a_2^2 + a_3^2}}$$

$$\cos \gamma = \frac{a_3}{|\mathbf{A}|} = \frac{a_3}{\sqrt{a_1^2 + a_2^2 + a_3^2}}.$$

Thus, for the vector $\mathbf{A} = a_1\mathbf{i} + a_2\mathbf{j} + a_3\mathbf{k}$, the direction cosines are given by

$$\cos \alpha = \frac{a_1}{\sqrt{a_1^2 + a_2^2 + a_3^2}},$$

$$\cos \beta = \frac{a_2}{\sqrt{a_1^2 + a_2^2 + a_3^2}},$$

$$\text{and} \quad \cos \gamma = \frac{a_3}{\sqrt{a_1^2 + a_2^2 + a_3^2}}.$$

and hence the direction cosines of $\mathbf{A}$ are proportional to a_1, a_2 and a_3, respectively. The direction cosines of an arbitrary vector are identical with the direction cosines of the unit vector pointing in its direction. As the sum of the squares of the coordinates of a unit vector is one, the sum of the squares of the direction cosines of any direction is also one:

$$\cos^2 \alpha + \cos^2 \beta + \cos^2 \gamma = 1.$$

The direction cosines of the vector $\mathbf{A} = 2\mathbf{i} - \mathbf{j} + \mathbf{k}$ are the following:

$$\cos \alpha = \frac{2}{\sqrt{4+1+1}} = \frac{2}{\sqrt{6}}$$

$$\cos \beta = \frac{-1}{\sqrt{4+1+1}} = \frac{-1}{\sqrt{6}}$$

$$\cos \gamma = \frac{1}{\sqrt{4+1+1}} = \frac{1}{\sqrt{6}}.$$

2.8 Lines and Planes Using Vector Algebra

2.8.1 Lines

A straight line can be determined by one of its points and its direction (i.e., the direction vector that is parallel to the straight line) or by two of its points.

1. **Equation of a Straight Line Determined by a Point and a Direction**

   The vector equation of a straight line passing through the point $P_0(x_0, y_0, z_0)$ and parallel to the vector $\mathbf{C} = a\mathbf{i} + b\mathbf{j} + c\mathbf{k}$ can be determined as follows:

   DEFINITION 2.9   *If we denote a moving point of the straight line by $P(x, y, z)$, then for the points of the straight line passing through the point P_0 in the direction $\mathbf{C}$ (and only for these points) the following equality holds:*

   $$\vec{P_0P} \times \mathbf{C} = \mathbf{0}$$

   If we denote the radius of an arbitrary point of the straight line by $\mathbf{r}$, and the radius vector leading up to P_0 by $\mathbf{r}_0$, then

   $$\vec{P_0P} = \mathbf{r} - \mathbf{r}_0$$

   Thus, the vector equation of the straight line is given by:

$$(\mathbf{r} - \mathbf{r}_0) \times \mathbf{V} = \mathbf{0}.$$

Expressed by coordinates:

$$\mathbf{r} - \mathbf{r}_0 = (x - x_0)\mathbf{i} + (y - y_0)\mathbf{j} + (z - z_0)\mathbf{k}$$
$$\text{and} \quad \mathbf{C} = a\mathbf{i} + b\mathbf{j} + c\mathbf{k}$$

thus,

$$(\mathbf{r} - \mathbf{r}_0) \times \mathbf{C} = \begin{vmatrix} \mathbf{i} & \mathbf{j} & \mathbf{k} \\ (x - x_0) & (y - y_0) & (z - z_0) \\ a & b & c \end{vmatrix}$$
$$= [(y - y_0)c - (z - z_0)b]\mathbf{i} - [(x - x_0)c - (z - z_0)a]\mathbf{j}$$
$$+ [(x - x_0)b - (y - y_0)a]\mathbf{k} = \mathbf{0}.$$

Hence, the vector equation of the straight line expressed by coordinates is:

$$[(y - y_0)c - (z - z_0)b]\mathbf{i} + [(z - z_0)a - (x - x_0)c]\mathbf{j}$$
$$+ [(x - x_0)b - (y - y_0)a]\mathbf{k} = \mathbf{0}.$$

Recall that a vector is zero if and only if all its coordinates are zero, i.e.,

$$(y - y_0)c - (z - z_0)b = 0$$
$$(z - z_0)a - (x - x_0)c = 0$$
$$(x - x_0)b - (y - y_0)a = 0$$

provided none of a, b, or c are zero. After reduction, these become:

$$\frac{x - x_0}{a} = \frac{y - y_0}{b} = \frac{z - z_0}{c}.$$

This is the scalar system for equations of the straight line. Among the three equations of this system only two are independent. If any two of them hold, the third one follows.

Example 2.8

Find the equation or system of equations of the straight line passing through the point $P_0(-2, 3, 5)$ and running in the direction $\mathbf{V} = 4\mathbf{i} - \mathbf{j} + 6\mathbf{k}$. $\square$

SOLUTION   The vector equation of the straight line is:

$$(\mathbf{r} - \mathbf{r}_0) \times \mathbf{C} = \begin{vmatrix} \mathbf{i} & \mathbf{j} & \mathbf{k} \\ x+2 & y-3 & z-5 \\ 4 & -1 & 6 \end{vmatrix} = \mathbf{0}$$

The scalar system of equations of the straight line is

$$\frac{x+2}{4} = \frac{y-3}{-1} = \frac{z-5}{6}.$$

$\blacksquare$

2. Equation of the Straight Line Determined by Two Points

The point $P(x, y, z)$ is on the straight line passing through the points $P_1(x_1, y_1, z_1)$ and $P_2(x_2, y_2, z_2)$ if and only if the vector $\mathbf{P_1P}$ is parallel to the vector $\mathbf{P_1P_2}$ determining the direction of the straight line. As we can determine the direction of the straight line from two points, the equation of the straight line determined by two points can be reduced to the equation of the straight line given by a point and a direction. The vector equation of the line passing through the points $P_1(x_1, y_1, z_1)$ and $P_2(x_2, y_2, z_2)$ is given by:

$$(\mathbf{r} - \mathbf{r}_i) \times \mathbf{P_1P_2} = \mathbf{0}$$

where $\mathbf{r}$ is the radius if the vector of an arbitrary point of the straight line, $\mathbf{r}_i$ the radius vector of the point P_i, $i = 1$, or 2. Expressed in coordinate form, this is

$$\begin{vmatrix} \mathbf{i} & \mathbf{j} & \mathbf{k} \\ x-x_1 & y-y_1 & z-z_1 \\ x_2-x_1 & y_2-y_1 & z_2-z_1 \end{vmatrix} = \mathbf{0}.$$

Thus the equation of the straight line passing through two points is given by

$$\frac{x-x_1}{x_2-x_1} = \frac{y-y_1}{y_2-y_1} = \frac{z-z_1}{z_2-z_1}.$$

2.8.2   The Equation of a Plane

A plane in space can be characterized in various ways:

1. by two intersecting lines

2. by a line and a point on the line

3. by three non-collinear points

For our present purpose, we use the fact that, given a point $P(x, y, z)$ and a vector (a, b, c), there is one and only one plane through P that is perpendicular to (a, b, c).

To find the equation of the plane we consider a point $Q(x, y, z)$ which is any other point on the plane. Then the vector $\mathbf{PQ} = (x - x_1, y - y_1, z - z_1)$ lies in the plane. But $\mathbf{PQ}$ must be perpendicular to the given vector (a, b, c) and hence by the vector dot product we have,

$$(a, b, c) \cdot (x - x_1, y - y_1, z - z_1) = 0$$

which is equivalent to

$$a(x - x_1) + b(y - y_1) + c(z - z_1) = 0.$$

This is the required equation of a plane passing through a point and perpendicular (or normal) to a given vector.

Example 2.9
Find the equation of the plane through the point (4,-3,5) and normal to the vector (4,-8,3).   ⬜

SOLUTION
The equation of the plane is

$$4(x - 4) - 8(y + 3) + 3(z - 5) = 0$$

or $4x - 8y + 3z = 55$.   ∎

2.9   Partial Derivatives

Before proceeding to the next chapter, it is necessary to briefly discuss and introduce the concept of partial derivatives. From elementary calculus, we are familiar with functions of a single variable. In real life, physical quantities usually depend on two or more variables. Fortunately, we merely need to extend the ideas of single-variable calculus to functions of several variables.

The temperature T at a point on the surface of the earth at any given time depends on the longitude x and latitude y of the point. We can think of T as being a function of the two variables x and y, or as a function of the pair (x, y). This functional dependence is indicated by writing $T = f(x, y)$. The volume V of a circular cylinder depends on its radius r and height h. We know from elementary calculus that $V = \pi r^2 h$. Here V is said to be a function of r and h, and we write $V(r, h) = \pi r^2 h$. As we have observed in the two examples, we often write $z = f(x, y)$ to make explicit the value taken on by f at the general point (x, y). The variables x and y are **independent variables** and z is the **dependent variable**. One can compare this with the single variable notation $y = f(x)$. Functions of two or more variables are also known as multi-variable functions.

Suppose that f is a function of two variables x and y. If we let only x vary while keeping y fixed, say $y = b$, then we are only considering a function of a single variable x, namely $z = f(x, b)$. If z has a derivative at a, then we call it the partial derivative of f with respect to x at (a, b). Similarly, the partial derivative of f with respect to y at (a, b) is obtained by keeping x fixed $(x = a)$ and finding the ordinary derivative at b of the function $z = f(a, y)$.

The same limiting notation for single variable functions is applicable to multi-variable functions. This leads us to the following definition for partial differentiation.

DEFINITION 2.10   *The partial derivative of a function $f(x, y)$ with respect to x is*

$$\frac{\partial f}{\partial x} = \lim_{\Delta x \to 0} \frac{f(x + \Delta x, y) - f(x, y)}{\Delta x} \tag{2.13}$$

and the partial derivative with respect to y is

$$\frac{\partial f}{\partial y} = \lim_{\Delta y \to 0} \frac{f(x, y + \Delta y) - f(x, y)}{\Delta y}. \tag{2.14}$$

It can be clearly seen from (2.13) that the partial derivative of $f(x, y)$ with respect to x is simply the ordinary derivative of $f(x, y)$ with respect to x, where y is considered a constant. Similarly, $\partial f / \partial y$ is the ordinary derivative of $f(x, y)$ with respect to y, holding x constant. For partial derivatives of functions of more than two independent variables, we again permit only one variable to vary while holding all the others constant. As an example, the partial derivative of $f(x, y, z, t, \ldots)$ with respect to z is

$$\frac{\partial f}{\partial z} = \lim_{\Delta z \to 0} \frac{f(x, y, z + \Delta z, t, \ldots) - f(x, y, z, t, \ldots)}{\Delta z}.$$

Therefore, we differentiate with respect to z while treating $x, y, t, \ldots$ as constants. There are other commonly used notations for the partial derivative. If $z = f(x, y)$, we write

$$f_x(x, y) = f_x = \frac{\partial f}{\partial x} = \frac{\partial}{\partial x} f(x, y) = \frac{\partial z}{\partial x} = D_x f$$

$$f_y(x, y) = f_y = \frac{\partial f}{\partial y} = \frac{\partial}{\partial y} f(x, y) = \frac{\partial z}{\partial y} = D_y f.$$

Note that here we cannot interpret $\partial f / \partial x$ as a ratio of differentials.

Example 2.10
If $f(x, y) = x^3 + x^2 y^3 - 2y^2$, find $f_x(2, 1)$ and $f_y(2, 1)$.   ◻

SOLUTION   Holding y constant and differentiating partially with respect to x, we get

$$f_x(x, y) = 3x^2 + 2xy^3$$

and so

$$f_x(2, 1) = 3 \cdot 2^2 + 2 \cdot 2 \cdot 1^3 = 16.$$

Holding x constant and differentiating partially with respect to y, we get

$$f_y(x, y) = 3x^2 y^2 - 4y$$
$$f_y(2, 1) = 3 \cdot 2^2 \cdot 1^2 - 4 \cdot 1 = 8.$$

∎

Example 2.11
Find f_x, f_y and f_z if $f(x, y, z) = e^{xy} \ln z$.   ◻

SOLUTION   Holding y and z constant and differentiating partially with respect to x, we have

$$f_x = y e^{xy} \ln z.$$

Similarly,

$$f_y = xe^{xy}\ln z \quad \text{and} \quad f_z = \frac{e^{xy}}{z}.$$

∎

Example 2.12
Find $\partial z/\partial x$ and $\partial z/\partial y$ if $z = f(x,y) = x^2y^3 + e^{xy}$.   ▯

SOLUTION   For this function,

$$\frac{\partial z}{\partial x} = 2xy^3 + ye^{xy} \quad \text{and} \quad \frac{\partial z}{\partial y} = 3x^2y^2 + xe^{xy}.$$

∎

2.10   Iterated Partial Derivatives

If f is a function of two variables, then its partial derivatives f_x and f_y are also functions of two variables. We can, therefore, consider their partial derivatives $(f_x)_x$, $(f_x)_y$, $(f_y)_x$, $(f_y)_y$ which are called the **second partial derivatives** of f. If $z = f(x,y)$, we use the following notation for the second partial derivatives:

$$(f_x)_x = f_{xx} = \frac{\partial}{\partial x}\left(\frac{\partial f}{\partial x}\right) = \frac{\partial^2 f}{\partial x^2} = \frac{\partial^2 z}{\partial x^2}$$

$$(f_x)_y = f_{xy} = \frac{\partial}{\partial y}\left(\frac{\partial f}{\partial x}\right) = \frac{\partial^2 f}{\partial y \partial x} = \frac{\partial^2 z}{\partial y \partial x}$$

$$(f_y)_x = f_{yx} = \frac{\partial}{\partial x}\left(\frac{\partial f}{\partial y}\right) = \frac{\partial^2 f}{\partial x \partial y} = \frac{\partial^2 z}{\partial x \partial y}$$

$$(f_y)_y = f_{yy} = \frac{\partial}{\partial y}\left(\frac{\partial f}{\partial y}\right) = \frac{\partial^2 f}{\partial y^2} = \frac{\partial^2 z}{\partial y^2}.$$

The notation f_{xy} (or $\partial^2 f/\partial y \partial x$) means that we first differentiate with respect to x and then with respect to y as opposed to f_{yx} where the order is reversed.

Example 2.13
Find the second partial derivatives of

$$f(x, y) = x^3 + x^2 y^3 - 2y^2.$$

⬚

SOLUTION   In Example 2.10 we found that

$$f_x(x, y) = 3x^2 + 2xy^3 \quad f_y(x, y) = 3x^2 y^2 - 4y.$$

Therefore, we obtain

$$f_{xx} = \frac{\partial}{\partial x}(3x^2 + 2xy^3) = 6x + 2y^3$$

$$f_{xy} = \frac{\partial}{\partial y}(3x^2 + 2xy^3) = 6xy^2$$

$$f_{yx} = \frac{\partial}{\partial x}(3x^2 y^2 - 4y) = 6xy^2$$

$$f_{yy} = \frac{\partial}{\partial y}(3x^2 y^2 - 4y) = 6x^2 y - 4.$$

∎

From this example we can note that $f_{xy} = f_{yx}$. This is not merely coincidence. It has been proven that the mixed partial derivatives f_{xy} and f_{yx} are equal for most functions that one meets in practice. Most texts on advanced calculus cite the French mathematician, Alexis Clairaut's (1713–1765) theorem giving the conditions under which we can assert that $f_{xy} = f_{yx}$.

Example 2.14
The general form of the wave equation is

$$u_{tt} = c^2 u_{xx}.$$

Verify that $u = f(x - ct) + g(x + ct)$ is a solution of the above equation where f and g are arbitrary regressive and progressive wave functions, respectively.
⬚

SOLUTION   Given that $u = f(x - ct) + g(x + ct)$ is a solution, we need to find the partial derivatives u_{tt} and u_{xx} first. Therefore, we have

$$u_t = -cf'(x - ct) + cg(x + ct)$$

so that $u_{tt} = c^2 f''(x - ct) + c^2 g(x + ct),$ and also,

$$u_x = f'(x - ct) + g'(x + ct), \quad \text{which leads to}$$

$$u_{xx} = f''(x - ct) + g''(x + ct) \quad \text{one can see that}$$

$$u_{tt} = c^2 (f''(x - ct) + g''(x + ct)), \quad \text{which equals}$$

$$= c^2 u_{xx}.$$

∎

Example 2.15

The partial differential equation $u_t + u_{xxx} + uu_x = 0$, called the **Korteweg–de Vries equation** (or KdV equation, for short), describes the motion of water waves in a shallow channel. (See Rahman's delightful book on water waves, published by Oxford University Press[12]).

(a) Show that for any positive constant c, the function

$$u(x,t) = 3c \ \mathrm{sech}^2 [\frac{1}{2}(x - ct)\sqrt{c}]$$

is a solution of the Korteweg–de Vries equation. (This solution represents a travelling *hump* of water in the channel and is called a **soliton**.)[1]

(b) How do the shape and speed of the soliton depend on c?

⬚

SOLUTION   (a) We compute u_t, u_x, u_{xx} and u_{xxx} using the chain rule and the differentiation formula $\frac{d}{dx}\mathrm{sech}\,x = -\mathrm{sech}\,x \tanh x$ from single-variable calculus. Then by letting $\alpha = (x - ct)\sqrt{c}/2$, we have

$$u = 3c\,\mathrm{sech}^2\alpha$$

$$u_t = 6c\ \mathrm{sech}\,\alpha\frac{\partial}{\partial t}\mathrm{sech}\,\alpha = -6c\ \mathrm{sech}^2\alpha\tanh\alpha\frac{\partial\alpha}{\partial t}$$

$$= 3c^{5/2}\ \mathrm{sech}^2\alpha\tanh\alpha = c^{3/2}u\tanh\alpha.$$

---

[1]Solitons were first observed, by J. Scott Russell around 1840, in barge canals near Edinburgh. He reported his results in *Trans. Royal Society of Edinburgh*, **14** (1840): 47–109.

Also,

$$u_x = -6c \ \text{sech}^2\alpha \tanh\alpha \frac{\partial\alpha}{\partial x}$$

$$= -3c^{3/2} \ \text{sech}^2\alpha \tanh\alpha = -\sqrt{c}u\tanh\alpha,$$

$$cu_x = -3c^{5/2}\text{sech}^2\alpha\tanh\alpha$$

$$= -u_t$$

so that $u_t = -cu_x$ and this leads to

$$u_{xx} = -\sqrt{c}(\tanh\alpha)u_x - \sqrt{c}u\text{sech}^2\alpha\frac{\sqrt{c}}{2}$$

$$= -\sqrt{c}u_x\tanh\alpha - \frac{c}{2}u\text{sech}^2\alpha$$

$$= -\sqrt{c}u_x\tanh\alpha - \frac{u^2}{6}.$$

Now using the value of u_x, u_{xx} can be obtained as

$$u_{xx} = cu - \frac{u^2}{2}.$$

And hence,

$$u_{xxx} = cu_x - uu_x.$$

Therefore, we have

$$u_t + u_{xxx} + uu_x = u_t + u_{xxx} + (cu_x - u_{xxx}) = u_t + cu_x = 0.$$

(b) The speed of the soliton is c, because the graph in Figure 2.10 (a) moves c units in x in one unit of t. The shape at time $t = 1$ is shown in Figure 2.10 (b) for $c = 1$. ∎

Exercises

1. Show by two methods that the vectors $\mathbf{A} = 2\mathbf{i}-3\mathbf{j}-\mathbf{k}$, $\mathbf{B} = -6\mathbf{i}+9\mathbf{j}+3\mathbf{k}$ are parallel.

2. Find a unit vector perpendicular to the vectors

$$\mathbf{A} = \mathbf{i} - \mathbf{j} + \mathbf{k}, \quad \text{and} \quad \mathbf{B} = \mathbf{i} + \mathbf{j} - \mathbf{k}.$$

3. Find the cosine of the angle between the two vectors $\mathbf{A} = 2\mathbf{i} - 3\mathbf{j} + \mathbf{k}$ and $\mathbf{B} = 3\mathbf{i} - \mathbf{j} - 2\mathbf{k}$.

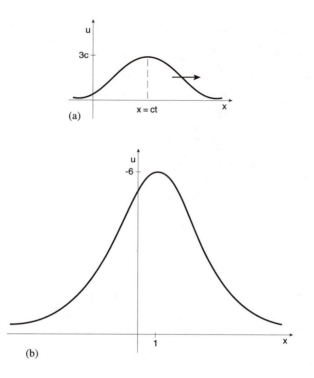

FIGURE 2.11
The soliton: (a) general; (b) for $c = 1$, $t = 1$

4. Find the unit vector which is parallel to the plane of $i + j - 2k$ and $3i - 2j + k$ and perpendicular to $2i + 2j - k$.

5. Show that $(A \times B) \cdot (B \times C) \times (C \times A) = [ABC]^2$.

6. Determine whether the given vectors are orthogonal, parallel or neither

   (a) $A = 2i - 4j$, $B = -i + 2j$
   (b) $A = 2i + j$, $B = 4i + 2j$
   (c) $A = 2i + 8j$, $B = -i + 2j$.

7. Find a unit vector that is orthogonal to $i + 3j$.

8. Find the scalar and vector projections of B onto A

   (a) $A = 2i + 3j$, $B = 4i + j$
   (b) $A = 3i - j$, $B = 2i + 3j$
   (c) $A = 4i + 2j$, $B = i + j$

(d) $\mathbf{A} = -\mathbf{i} - 2\mathbf{j}$, $\mathbf{B} = 3\mathbf{i} + 3\mathbf{j}$

(e) $\mathbf{A} = \mathbf{i}$, $\mathbf{B} = \mathbf{i} - \mathbf{j}$

(f) $\mathbf{A} = 2\mathbf{i} - 3\mathbf{j}$, $\mathbf{B} = \mathbf{i} + 6\mathbf{j}$.

9. (a) Show

$$\mathbf{A} \times (\mathbf{B} \times \mathbf{C}) + \mathbf{B} \times (\mathbf{C} \times \mathbf{A}) + \mathbf{C} \times (\mathbf{A} \times \mathbf{B}) = \mathbf{0}.$$

(b) Show that if four vectors $\mathbf{A}$, $\mathbf{B}$, $\mathbf{C}$, $\mathbf{D}$ are coplanar then $(\mathbf{A} \times \mathbf{B}) \times (\mathbf{C} \times \mathbf{D}) = \mathbf{0}$.

10. (a) Find the distance from the point $(3,7)$ to the line $y = 2x$.

(b) Find the distance from the point $(-2, 3)$ to the line $3x + 4y + 5 = 0$.

11. Find the distance between the parallel lines $y = -4x + 3$ and $y = -4x + 12$.

12. Show that the volume of the parallelepiped having $\mathbf{A} + \mathbf{B}$, $\mathbf{B} + \mathbf{C}$, and $\mathbf{C} + \mathbf{A}$ as concurrent edges is twice the volume of the parallelepiped having $\mathbf{A}$, $\mathbf{B}$, and $\mathbf{C}$ as concurrent edges.

13. (a) Use a scalar projection to show that the distance from a point $P_1(x_1, y_1)$ to the line $ax + by + c = 0$ is

$$\frac{|ax_1 + by_1 + c|}{\sqrt{a^2 + b^2}}.$$

(b) Use this formula to find the distance from the point $(-2, 3)$ to the line $3x - 4y + 5 = 0$.

14. Find the area of a parallelogram if the vertices are $(1, 1)$, $(4, -2)$, $(9, 3)$, $(12, 0)$.

15. Find the area of a parallelogram if the vertices are $(1, 1, 1)$, $(4, 4, 4)$, $(8, -3, 14)$, $(11, 0, 17)$.

16. Find a normal vector to the plane through the points $(1, 3, 0)$, $(2, 0, 8)$, $(0, 2, 2)$ and from it find an equation for the plane.

17. Find the volume of the parallelepiped if the edge vectors are:

(a) $\mathbf{i} + \mathbf{j}$, $-2\mathbf{i} + 2\mathbf{k}$, $-2\mathbf{i} - 3\mathbf{k}$

(b) $(4, 9, -1)$, $(2, 0, 0)$, $(5, -4, 2)$.

18. Find the length of each of the following vectors and the angle it makes with the positive x-axis:

(a) $\mathbf{i} + \mathbf{j}$

(b) $-\mathbf{i} + \mathbf{j}$

(c) $\mathbf{i} + \sqrt{3}\mathbf{j}$

(d) $\mathbf{i} - \sqrt{3}\mathbf{j}$.

19. Write the equation of the plane

   (a) passing through $P_0(1, 2, 3)$ and parallel to $3x - 2y + 4z - 5 = 0$

   (b) passing through $P_0(1, 2, 3)$ and $P_1(3, -2, 1)$, and perpendicular to the plane $3x - 2y + 4z - 5 = 0$

   (c) through $P_0(1, 2, 3)$, $P_1(3, -2, 1)$ and $P_2(5, 0, -4)$. Let $P(x, y, z)$ be a general point in the required plane.

20. Prove that

$$(\mathbf{A} \times \mathbf{B}) \cdot (\mathbf{C} \times \mathbf{D}) = (\mathbf{A} \cdot \mathbf{C})(\mathbf{B} \cdot \mathbf{D}) - (\mathbf{A} \cdot \mathbf{D})(\mathbf{B} \cdot \mathbf{C}).$$

21. In complex variable theory, two functions $u(x, y)$ and $v(x, y)$ are said to be *harmonic conjugates* in a region R if in R they satisfy the Cauchy–Riemann equations

$$\frac{\partial u}{\partial x} = \frac{\partial v}{\partial y}, \quad \frac{\partial v}{\partial x} = -\frac{\partial u}{\partial y}.$$

   Show that the following pairs of functions are harmonic conjugates:

   (a) $u(x, y) = (x^2 + x + y^2)/(x^2 + y^2)$, $v(x, y) = -y/(x^2 + y^2)$

   (b) $u(x, y) = e^x(x \cos y - y \sin y)$, $v(x, y) = e^x(x \sin y + y \cos y)$.

22. If r and θ are polar coordinates, then the Cauchy–Riemann equations from the previous question for functions $u(r, \theta)$ and $v(r, \theta)$ take the form

$$\frac{\partial u}{\partial r} = \frac{1}{r}\frac{\partial v}{\partial \theta}, \quad \frac{1}{r}\frac{\partial u}{\partial \theta} = -\frac{\partial v}{\partial r}, \quad r \neq 0.$$

   Show that the following pairs of functions satisfy these equations:

   (a) $u(r, \theta) = (r^2 + r \cos \theta)/(1 + r^2 + 2r \cos \theta)$, $v(r, \theta) = r \sin \theta/(1 + r^2 + 2r \cos \theta)$

   (b) $u(r, \theta) = \sqrt{r} \cos(\theta/2)$, $v(r, \theta) = \sqrt{r} \sin(\theta/2)$

   (c) $u(r, \theta) = \ln r$, $v(r, \theta) = \theta$.

23. Verify that $u = e^{-\alpha t} \sin x$ is a solution of the heat equation

$$u_t = \alpha u_{xx}.$$

24. Verify that $u = \ln r$, where $r = \sqrt{(x - x_0)^2 + (y - y_0)^2}$ is a solution of the two-dimensional Laplace equation

$$u_{xx} + u_{yy} = 0.$$

25. Verify that $u = \frac{1}{r}$, where

$$r = \sqrt{(x - x_0)^2 + (y - y_0)^2 + (z - z_0)^2}$$

is a solution of the three-dimensional Laplace equation

$$u_{xx} + u_{yy} + u_{zz} = 0.$$

Chapter 3

Vector Functions of One Variable

When solving problems in physics and mathematics we must often examine vector quantities depending on scalar or vector variables, or scalar quantities depending on vectors. For example, the speed of a moving point depends on the time, i.e., on a scalar variable; the intensity of electric field, which is a scalar quantity, depends on the radius vector of the point, that is to say on a vector variable. The branch of mathematics that deals with variable vector quantities, vector functions, is called vector analysis. There are 3 types of functions in vector analysis:

1. **Vector–scalar** functions: the independent variable is a scalar, the dependent variable a vector.

2. **Scalar–vector** functions: the independent variable is a vector, the dependent variable is a scalar.

3. **Vector–vector** functions: both the independent and dependent variable are vectors.

If t is a scalar variable, and if to each value of t in some interval there corresponds a value of a vector $\mathbf{V}$, we say that $\mathbf{V}$ is a **vector function** of t. We use the letter t to denote the independent variable because it represents time in most applications of vector functions. Because the component of a vector in any direction is known whenever the vector itself is known, it follows that if $\mathbf{V}$ is a function of t, so too are its components in the directions of the unit vectors $\mathbf{i}$, $\mathbf{j}$ and $\mathbf{k}$. We denote the vector function with one variable by $\mathbf{V}(t)$ where the independent variable t is a scalar quantity, the dependent variable $\mathbf{V}$ is a vector quantity. Any n-dimensional vector function $\mathbf{V}(t)$ can be given by means of n scalar functions. The three-dimensional case is dealt with in this analysis.

Let the coordinates of the vector $\mathbf{V}$ be V_1, V_2, V_3. Some value of the variable t determines the vector $\mathbf{V}(t)$ and thus determines its coordinates, V_1, V_2, V_3. If we denote the function relations that determine the dependence of the coordinates V_1, V_2, V_3 upon t by $V_1(t)$, $V_2(t)$, $V_3(t)$, then we can write down the vector function $\mathbf{V}(t)$ in the following form:

$$\mathbf{V}(t) = V_1(t)\mathbf{i} + V_2(t)\mathbf{j} + V_3(t)\mathbf{k}. \tag{3.1}$$

In particular, we say that $\mathbf{V}(t)$ is continuous if and only if the three scalar functions $V_1(t)$, $V_2(t)$, and $V_3(t)$ are continuous. The scalar variable of the vector function is called the **parameter** of the space curve. Thus, the vector function

$$\mathbf{V}(t) = V_1(t)\mathbf{i} + V_2(t)\mathbf{j} + V_3(t)\mathbf{k}$$

written by means of coordinates:

$$\mathbf{V}(t) = \begin{cases} V_1 = V_1(t) \\ V_2 = V_2(t) \\ V_3 = V_3(t) \end{cases}$$

is at the same time a parametric system of equations of the space curve. Note that if one of the coordinates of the vector functions $\mathbf{V} = V_1(t)\mathbf{i} + V_2(t)\mathbf{j} + V_3(t)\mathbf{k}$ is a constant, the curve is a plane lying – according to the constant coordinate – in a plane parallel to one of the coordinate planes.

3.1   Vector Differentiation

Suppose the independent variable t of a vector function $\mathbf{V}(t)$ changes by an amount Δt. Then the function will change both in magnitude and direction. Corresponding to the scalar increment Δt we have the vector increment

$$\begin{aligned} \Delta\mathbf{V} &= \mathbf{V}(t + \Delta t) - \mathbf{V}(t) \\ &= [V_1(t + \Delta t)\mathbf{i} + V_2(t + \Delta t)\mathbf{j} + V_3(t + \Delta t)\mathbf{k}] \\ &\quad - [V_1(t)\mathbf{i} + V_2(t)\mathbf{j} + V_3(t)\mathbf{k}] \\ &= \Delta V_1(t)\mathbf{i} + \Delta V_2(t)\mathbf{j} + \Delta V_3(t)\mathbf{k}. \end{aligned} \tag{3.2}$$

If we now divide $\Delta\mathbf{V}$ by Δt and take the limit as $\Delta t \to 0$, we obtain the new vector known as the derivative of a vector function $\mathbf{V}(t)$ with respect to t

$$\frac{d\mathbf{V}}{dt} = \lim_{\Delta t \to 0} \frac{\mathbf{V}(t + \Delta t) - \mathbf{V}(t)}{\Delta t} = \lim_{\Delta t \to 0} \frac{\Delta\mathbf{V}}{\Delta t}. \tag{3.3}$$

Using (3.2), this becomes

$$\frac{d\mathbf{V}}{dt} = \lim_{\Delta \to 0} \frac{\Delta V_1(t)}{\Delta t}\mathbf{i} + \lim_{\Delta t \to 0} \frac{\Delta V_2(t)}{\Delta t}\mathbf{j} + \lim_{\Delta t \to 0} \frac{\Delta V_3(t)}{\Delta t}\mathbf{k}$$
$$= \frac{dV_1(t)}{dt}\mathbf{i} + \frac{dV_2(t)}{dt}\mathbf{j} + \frac{dV_3(t)}{dt}\mathbf{k}. \tag{3.4}$$

From (3.4) we are motivated to define the **differential of a vector function** $\mathbf{V}(t)$ to be

$$d\mathbf{V} = dV_1\mathbf{i} + dV_2\mathbf{j} + dV_3\mathbf{k}. \tag{3.5}$$

The notation $\mathbf{V}'$ is often used to signify the derivative when the variable of differentiation is clear from the context. It is also common to place a dot above the vector to indicate total differentiation with respect to time t, i.e.,

$$\dot{\mathbf{V}} = \frac{d\mathbf{V}}{dt}.$$

The expression above implies that, in terms of components with respect to a given Cartesian coordinate system, $\mathbf{V}(t)$ is differentiable at a point t if and only if its three components are differentiable at t. The derivative $\dot{\mathbf{V}}(t)$ is obtained by differentiating each component separately. For the position vector

$$\mathbf{R} = x\mathbf{i} + y\mathbf{j} + z\mathbf{k}$$

drawn from the origin to the point (x, y, z), we have

$$d\mathbf{R} = dx\mathbf{i} + dy\mathbf{j} + dz\mathbf{k} \quad \text{or}$$
$$\frac{d\mathbf{R}}{dt} = \frac{dx}{dt}\mathbf{i} + \frac{dy}{dt}\mathbf{j} + \frac{dz}{dt}\mathbf{k}.$$

As can be seen from the expression above, on a practical level the derivative of a vector function is obtained by simply differentiating its component functions. When the components of a vector function $\mathbf{R}$ have continuous first derivatives and $\mathbf{R}' \neq 0$ for all t in the open interval (a, b), then $\mathbf{R}$ is said to be a **smooth function** and the curve C traced by $\mathbf{R}$ is called a smooth curve. We shall denote the absolute value of $\mathbf{R}$ by r, i.e., $|\mathbf{R}| = r$.

3.2 Geometric Interpretation of R′

If the vector $\mathbf{R}'$ is not $\mathbf{0}$ at a point P, then it may be drawn tangent to the curve at P. As seen in Figure 3.1 (a) and (b), the vectors

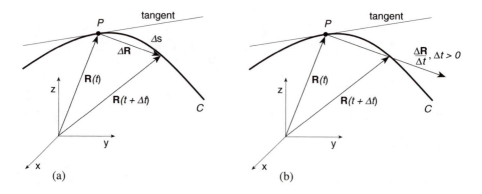

FIGURE 3.1
Geometrical interpretation of vector function

$$\Delta \mathbf{R} = \mathbf{R}(t + \Delta t) - \mathbf{R}(t) \quad \text{and} \quad \frac{\Delta \mathbf{R}}{\Delta t} = \frac{1}{\Delta t}[\mathbf{R}(t + \Delta t) - \mathbf{R}(t)]$$

are parallel. If we assume $\lim_{\Delta t \to 0} \frac{\Delta \mathbf{R}}{\Delta t}$ exists, it is reasonable to conclude that as $\Delta \to 0$, $\mathbf{R}(t)$ and $\mathbf{R}(t + \Delta t)$ become close and, as a consequence, the limiting position of the vector $\frac{\Delta \mathbf{R}}{\Delta t}$ is the tangent line at P. In fact the tangent line at P is defined as that line through P parallel to $\mathbf{R}'(t)$.

From the definition of the derivative of a vector function of one variable, it follows that sums, differences and products of vectors can be differentiated by formulae similar to those of ordinary calculus provided that the proper order of factors is maintained wherever the order is significant. Specifically, we have

$$\frac{d(\mathbf{U} \pm \mathbf{V})}{dt} = \frac{d\mathbf{U}}{dt} \pm \frac{d\mathbf{V}}{dt} \tag{3.6}$$

$$\frac{d(\phi\mathbf{V})}{dt} = \frac{d\phi}{dt}\mathbf{V} + \phi\frac{d\mathbf{V}}{dt} \tag{3.7}$$

$$\frac{d(\mathbf{U} \cdot \mathbf{V})}{dt} = \frac{d\mathbf{U}}{dt} \cdot \mathbf{V} + \mathbf{U} \cdot \frac{d\mathbf{V}}{dt} \tag{3.8}$$

$$\frac{d(\mathbf{U} \times \mathbf{V})}{dt} = \frac{d\mathbf{U}}{dt} \times \mathbf{V} + \mathbf{U} \times \frac{d\mathbf{V}}{dt} \tag{3.9}$$

$$\frac{d[\mathbf{U} \times (\mathbf{V} \times \mathbf{W})]}{dt} = \frac{d\mathbf{U}}{dt} \times (\mathbf{V} \times \mathbf{W}) + \mathbf{U} \times \left\{\frac{d\mathbf{V}}{dt} \times \mathbf{W}\right\}$$

$$+ \mathbf{U} \times \left\{\mathbf{V} \times \frac{d\mathbf{W}}{dt}\right\}. \tag{3.10}$$

Because every vector equation implies three simultaneous scalar equations, it follows from the definition of vector differentiation that the integral of a vector function of one variable is found by integrating each component, provided only that the constant of integration is an arbitrary constant vector. Therefore, if

$$\mathbf{V}(t) = \mathbf{V}_1(t)\mathbf{i} + \mathbf{V}_2(t)\mathbf{j} + \mathbf{V}_3(t)\mathbf{k} \quad \text{then}$$

$$\int \mathbf{V}(t)dt = \mathbf{i} \int \mathbf{V}_1(t)dt + \mathbf{j} \int \mathbf{V}_2(t)dt + \mathbf{k} \int \mathbf{V}_3(t)dt + \mathbf{C}$$

where the constant vector is $\mathbf{C} = C_1\mathbf{i} + C_2\mathbf{j} + C_3\mathbf{k}$. When $\Delta\mathbf{R}$ tends to zero, the elementary arc length is given by

$$\Delta s = |\Delta\mathbf{R}|.$$

If s and $\mathbf{R}$ are functions of the independent time variable t, and as $\Delta t \to 0$, then

$$\frac{ds}{dt} = \left|\frac{d\mathbf{R}}{dt}\right| = \sqrt{\frac{d\mathbf{R}}{dt} \cdot \frac{d\mathbf{R}}{dt}}.$$

Therefore, the total arc length can be obtained from

$$s = \int_{a(t)}^{b(t)} \sqrt{\left(\frac{dx}{dt}\right)^2 + \left(\frac{dy}{dt}\right)^2 + \left(\frac{dz}{dt}\right)^2} \, dt.$$

3.3   Higher-Order Derivatives

Higher-order derivatives of a vector function are also obtained by differentiating its components. In the case of the second derivative, we have

$$\mathbf{R}'' = f''(t)\mathbf{i} + g''(t)\mathbf{j} + h''(t)\mathbf{k}$$

where $f(t)$, $g(t)$, and $g(t)$ are the components of $\mathbf{R}(t)$ with first and second derivatives.

As an example, consider $\mathbf{R}(t) = (t^3 - 2t^2)\mathbf{i} + 4t\mathbf{j} + e^{-t}\mathbf{k}$. It follows that

$$\mathbf{R}'(t) = (3t^2 - 4t)\mathbf{i} + 4\mathbf{j} - e^{-t}\mathbf{k}$$
$$\text{and} \quad \mathbf{R}'' = (6t - 4)\mathbf{i} + e^{-t}\mathbf{k}.$$

Also, as an example, if $\mathbf{R}(t) = 6t^2\mathbf{i} + 4e^{-2t}\mathbf{j} + 8\cos 4t\mathbf{k}$ then

$$\int \mathbf{R}(t)dt = \left(\int 6t^2 dt \right) \mathbf{i} + \left(\int 4e^{-2t} dt \right) \mathbf{j} + \left(\int 8\cos 4t dt \right) \mathbf{k}$$
$$= (2t^3 + C_1)\mathbf{i} + (-2e^{-2t} + C_2)\mathbf{j} + (2\sin 4t + C_3)\mathbf{k}$$
$$= 2t^3\mathbf{i} - 2e^{-2t}\mathbf{j} + 2\sin 4t\mathbf{k} + \mathbf{C}$$

where the constant vector $\mathbf{C} = C_1\mathbf{i} + C_2\mathbf{j} + C_3\mathbf{k}$.

3.4   Curves, Length and Arc Length

Curves in space are important in calculus and physics (i.e., as paths of moving bodies), and we show in the following sections that they constitute a major field of application of vector calculus. Their theory together with that of surfaces in space is called **differential geometry**. It plays a significant role in engineering design, geodesy, geography, space travel and relativity theory.

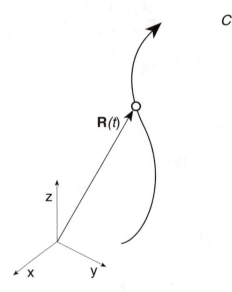

FIGURE 3.2
Parametric representation of a curve

A curve C in space can be represented by a vector function

$$\mathbf{R}(t) = (x(t), y(t), z(t)) = x(t)\mathbf{i} + y(t)\mathbf{j} + z(t)\mathbf{k} \qquad (3.11)$$

where x, y, z are Cartesian coordinates. This is known as a **parametric representation** of the curve, with t being the **parameter** of the representation. To each value t_0 there corresponds a point C with position vector $\mathbf{R}(t_0)$, i.e., with coordinates $x(t_0)$, $y(t_0)$, $z(t_0)$. The parameter t may represent time or some other physical quantity. In equation (3.11) the coordinates x, y, z play the same role — all three are dependent variables. We shall see that very often this is a great practical advantage, the key property of a parametric representation.

We will now define the length l of a curve. This length will be the limit of the lengths of broken lines of n chords (see Figure 3.3) with larger and larger n. To represent C, let us take $\mathbf{R}(t)$, $a \leq t \leq b$. For each $n = 1, 2, \ldots$ we subdivide (or partition) the interval $a \leq t \leq b$ by points

$$t_0(= a), t_1, \ldots, t_{n-1}, t_n(= b)$$

where $t_0 < t_1 < \ldots < t_n$. This gives a broken line of chords with endpoints $\mathbf{R}(t_0), \ldots, \mathbf{R}(t_n)$. This is done arbitrarily but so that the greatest $|\Delta t_m| = |t_m - t_{m-1}|$ approaches zero as $n \to \infty$. The lengths $l_1, l_2, \ldots$ of these lines of chords can be obtained from the Pythagorean theorem.

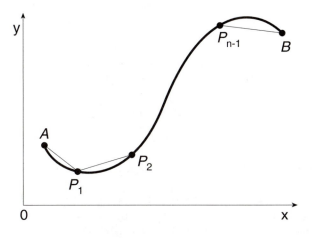

FIGURE 3.3
Length of a curve

If $\mathbf{R}(t)$ has a continuous derivative $\mathbf{R}'(t)$, it can be shown that the sequence $l_1, l_2, \ldots$ has a limit, which is independent of the particular choice of representation of C and of the choice of subdivisions. This limit is given by the integral

$$s = \int_a^b \sqrt{\mathbf{R}' \cdot \mathbf{R}'}\, dt \qquad (3.12)$$

where s is called the **length** of C and C is called **rectifiable**. The practical evaluation of the integral (3.12) will be generally difficult.

The length s of a curve is a constant — a positive number. In the expression (3.12), if we replace the fixed upper limit with the variable t, then the integral becomes a function of t. This function of t is commonly denoted by $s(t)$ and called the **arc length** of C. If $\mathbf{R}(t) = f(t)\mathbf{i} + g(t)\mathbf{j} + h(t)\mathbf{k}$ is a smooth function, then it can be shown that the length of the smooth curve traced by $\mathbf{R}$ is given by

$$s = \int_a^b \sqrt{(f'(t))^2 + (g'(t))^2 + (h'(t))^2}\,dt$$
$$= \int_a^b |\mathbf{R}'(t)|\,dt.$$

A curve in the plane or in space can be parameterized in terms of the arc length s. Consider the helix $\mathbf{R}(t) = 2\cos t\mathbf{i} + 2\sin t\mathbf{j} + t\mathbf{k}$, $t \geq 0$. Because $|\mathbf{R}'(t)| = \sqrt{5}$, it follows that the length of the curve from $\mathbf{R}(0)$ to an arbitrary point $\mathbf{R}(t)$ is

$$s = \int_0^t \sqrt{5}\,du = \sqrt{5}t$$

where we have used u as a dummy variable of integration. Using $t = s/\sqrt{5}$, we obtain a vector equation of the helix as a function of arc length:

$$\mathbf{R}(s) = 2\cos(s/\sqrt{5})\mathbf{i} + 2\sin(s/\sqrt{5})\mathbf{j} + (s/\sqrt{5})\mathbf{k}.$$

Parametric equations of the helix are then

$$f(s) = 2\cos(s/\sqrt{5}), \quad g(s) = 2\sin(s/\sqrt{5}), \quad h(s) = \frac{s}{\sqrt{5}}.$$

The derivative of a vector function $\mathbf{R}(t)$ with respect to the parameter t is a tangent vector to the curve traced by $\mathbf{R}$. However, if the curve is parameterized in terms of arc length s, then $\mathbf{R}'(s)$ is a unit tangent vector.

3.5   Motion on a Curve, Velocity and Acceleration

Suppose a body or a particle moves along a curve C so that its position at time t is given by

$$\mathbf{R}(t) = f(t)\mathbf{i} + g(t)\mathbf{j} + h(t)\mathbf{k}.$$

If f, g, and h have second derivatives, then the vectors

$$\mathbf{V}(t) = \mathbf{R}'(t) = f'(t)\mathbf{i} + g'(t)\mathbf{j} + h'(t)\mathbf{k} \qquad (3.13)$$
$$\mathbf{a}(t) = \mathbf{R}''(t) = f''(t)\mathbf{i} + g''(t)\mathbf{j} + h''(t)\mathbf{k} \qquad (3.14)$$

are called **velocity** and **acceleration** of the particle, respectively. The scalar function $|\mathbf{V}(t)|$ is the **speed** of the particle. Since the position vector $\mathbf{R}$ is usually defined by $\mathbf{R} = x\mathbf{i} + y\mathbf{j} + z\mathbf{k}$, we can write

$$|\mathbf{V}(t)| = \left|\frac{d\mathbf{R}}{dt}\right| = \sqrt{\left(\frac{dx}{dt}\right)^2 + \left(\frac{dy}{dt}\right)^2 + \left(\frac{dz}{dt}\right)^2}.$$

Speed is related to arc length s by $s'(t) = |\mathbf{V}(t)|$. In other words, arc length is given by

$$s = \int_{t_0}^{t_1} |\mathbf{V}(t)|\,dt.$$

Example 3.1
The position of a moving particle is given by $\mathbf{R}(t) = t^2\mathbf{i} + t\mathbf{j} + \frac{5}{2}t\mathbf{k}$. Graph the curve defined by $\mathbf{R}(t)$ and the vectors $\mathbf{V}(2)$ and $\mathbf{a}(2)$. ⬜

SOLUTION   Because $x = t^2$, $y = t, z = \frac{5}{2}t$, the path of the particle is above the parabola $x = y^2$. When $t = 2$ the position vector $\mathbf{R}(2) = 4\mathbf{i} + 2\mathbf{j} + 5\mathbf{k}$ indicates that the particle is at the point $P(4, 2, 5)$. Now

$$\mathbf{V}(t) = \mathbf{R}'(t) = 2t\mathbf{i} + \mathbf{j} + \frac{5}{2}\mathbf{k} \quad \text{and}$$
$$\mathbf{a}(t) = \mathbf{R}''(t) = 2\mathbf{i} \quad \text{so that}$$
$$\mathbf{V}(2) = 4\mathbf{i} + \mathbf{j} + \frac{5}{2}\mathbf{k} \quad \text{and} \quad \mathbf{a}(2) = 2\mathbf{i}.$$

■

These vectors are shown in Figure 3.4

3.6 Curvature, Components of Acceleration

Let C be a smooth curve in 2- or 3-space traced out by a vector function $\mathbf{R}(t)$. In this section we will discuss in greater detail the acceleration vector

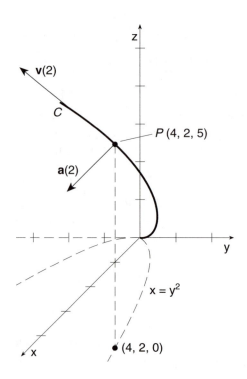

FIGURE 3.4
Curves defined by R(t), V(2) and a(2)

$\mathbf{a}(t) = \mathbf{R}''(t)$ introduced in the last section. Before doing this, we need to examine a scalar quantity called the **curvature** of a curve. It is known that $\mathbf{R}'(t)$ is a tangent vector to the curve C and consequently

$$\mathbf{T} = \frac{\mathbf{R}'(t)}{|\mathbf{R}'(t)|} \tag{3.15}$$

is a **unit tangent** vector. However, if C is parameterized by arc length s, then a unit tangent to the curve is also given by $\frac{d\mathbf{R}}{ds}$. The quantity $|\mathbf{R}'(t)|$ in (3.15) is related to arc length s by $\frac{ds}{dt} = |\mathbf{R}'(t)|$. Because the curve C is smooth, it is known that $\frac{ds}{dt} > 0$. Hence, by the chain rule, we have:

$$\frac{d\mathbf{R}}{dt} = \frac{d\mathbf{R}}{ds}\left(\frac{ds}{dt}\right)$$

we know $\quad\dfrac{ds}{dt} = V = \quad$ speed of the particle along C

and $\quad\dfrac{d\mathbf{R}}{ds} = \mathbf{T} = \quad$ unit vector tangent to C

Therefore, $\quad\dfrac{d\mathbf{R}}{dt} = v\mathbf{T}.$

Now $\mathbf{T}$ can be written as

$$\mathbf{T} = \frac{d\mathbf{R}}{ds} = \frac{d\mathbf{R}/dt}{ds/dt} = \frac{\mathbf{R}'(t)}{|\mathbf{R}'(t)|}.$$

Suppose C is as shown in Figure 3.5. As s increases, $\mathbf{T}$ moves along C, changing direction but not length (it is always of unit length). Along the portion of the curve between P_1 and P_2 the vector $\mathbf{T}$ varies little in direction; along the curve between P_2 and P_3, where C obviously bends more sharply, the change in the direction of the tangent $\mathbf{T}$ is more pronounced. We use the *rate* at which the unit vector $\mathbf{T}$ changes direction with respect to arc length as an indicator of a smooth curve C.

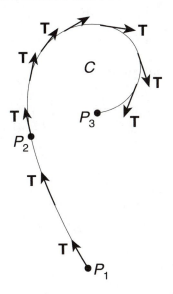

FIGURE 3.5
The tangent T as it moves along the arc

This leads to the following definition.

DEFINITION 3.1   *Let* $\mathbf{R}(t)$ *be a vector function defining a smooth curve* C. *If* s *is the arc length parameter and* $\mathbf{T} = \frac{d\mathbf{R}}{ds}$ *is the unit tangent vector, then the* **curvature** *of* C *at a point is defined as*

$$\kappa = \left|\frac{d\mathbf{T}}{ds}\right|. \tag{3.16}$$

The symbol κ in (3.16) is the Greek letter **kappa**. Now, because curves are generally not parameterized by arc length, it is convenient to express (3.16) in terms of a general parameter t. We can write $\frac{d\mathbf{T}}{ds}$ as

$$\frac{d\mathbf{T}}{ds} = \frac{d\mathbf{T}/dt}{ds/dt}$$

$$\text{So} \quad \left|\frac{d\mathbf{T}}{ds}\right| = \frac{|\mathbf{T}'(t)|}{|\mathbf{R}'(t)|}.$$

In other words, curvature is given by

$$\kappa = \frac{|\mathbf{T}'(t)|}{|\mathbf{R}'(t)|}. \tag{3.17}$$

Example 3.2
Find the curvature of a circle of radius a.   ⬜

SOLUTION   A circle can be described by the vector function $\mathbf{R}(t) = a\cos t\mathbf{i} + a\sin t\mathbf{j}$. Now from $\mathbf{R}'(t) = -a\sin t\mathbf{i} + a\cos t\mathbf{j}$ and $|\mathbf{R}'(t)| = a$, we get

$$\mathbf{T}(t) = \frac{\mathbf{R}'(t)}{|\mathbf{R}'(t)|} = -\sin t\mathbf{i} + \cos t\mathbf{j} \quad \text{and}$$
$$\mathbf{T}'(t) = -\cos t\mathbf{i} - \sin t\mathbf{j}.$$

Hence, the curvature is given by

$$\kappa = \frac{|\mathbf{T}'(t)|}{|\mathbf{R}'(t)|} = \frac{\sqrt{\cos^2 t + \sin^2 t}}{a} = \frac{1}{a}.$$

3.7 Curvature, Tangential and Normal Components of Acceleration

Suppose a particle moves in 2- or 3-space on a smooth curve C described by the vector function $\mathbf{R}(t)$. Then the velocity of the particle on C is $\mathbf{V}(t) = \mathbf{R}'(t)$, whereas its speed is $\frac{ds}{dt} = V = |\mathbf{V}(t)|$. Thus, we have

$$\mathbf{T} = \frac{\mathbf{V}(t)}{|\mathbf{V}(t)|} = \frac{\mathbf{V}(t)}{V}$$

so that $\mathbf{V}(t) = V\mathbf{T}$. Differentiating this last expression with respect to t gives acceleration:

$$\mathbf{a}(t) = V\frac{d\mathbf{T}}{dt} + \frac{dV}{dt}\mathbf{T}. \tag{3.18}$$

Furthermore, it follows from the differentiation of $\mathbf{T} \cdot \mathbf{T} = 1$ that $\mathbf{T} \cdot \frac{d\mathbf{T}}{dt} = 0$. Hence, at a point P on C the vectors $\mathbf{T}$ and $\frac{d\mathbf{T}}{dt}$ are orthogonal. If $|\frac{d\mathbf{T}}{dt}| \neq 0$, the vector

$$\mathbf{N} = \frac{d\mathbf{T}/dt}{|d\mathbf{T}/dt|} \tag{3.19}$$

is a unit normal to the curve C at P with direction given by $\frac{d\mathbf{T}}{dt}$. The vector $\mathbf{N}$ is also called the **principal normal**. But because curvature is $\kappa = \frac{|d\mathbf{T}/dt|}{V}$, it follows from (3.19) that $\frac{d\mathbf{T}}{dt} = \kappa V \mathbf{N}$. Thus, (3.18) becomes

$$\mathbf{a}(t) = \kappa V^2 \mathbf{N} + \frac{dV}{dt}\mathbf{T}. \tag{3.20}$$

Thus we have

$$\frac{d\mathbf{T}}{ds} = \frac{d\mathbf{T}/dt}{ds/dt} = \frac{\kappa V}{V}\mathbf{N} = \kappa\mathbf{N}. \tag{3.21}$$

A third unit vector defined by

$$\mathbf{B} = \mathbf{T} \times \mathbf{N}$$

is called the **binormal**. The three unit vectors, $\mathbf{T}$, $\mathbf{N}$ and $\mathbf{B}$ form a right-handed set of mutually orthogonal vectors called the **moving trihedral**. The plane of $\mathbf{T}$ and $\mathbf{N}$ is called the **osculating plane**,[1] the plane of $\mathbf{N}$ and $\mathbf{B}$ is said to be the **normal plane**, and the plane of $\mathbf{T}$ and $\mathbf{B}$ is the

---

[1]Literally, this means the "kissing" plane.

rectifying plane (see Figure 3.6) If we take the derivative of **B** with respect to s (arc length) we obtain the following

$$\frac{d\mathbf{B}}{ds} = \frac{d\mathbf{T}}{ds} \times \mathbf{N} + \mathbf{T} \times \frac{d\mathbf{N}}{ds}$$
$$= \kappa\mathbf{N} \times \mathbf{N} + \mathbf{T} \times \frac{d\mathbf{N}}{ds}$$
$$= \mathbf{T} \times \frac{d\mathbf{N}}{ds} = -\tau\mathbf{N}. \tag{3.22}$$

The quantity $\left|\frac{d\mathbf{B}}{ds}\right|$ is called the **torsion** of the curve at P. It is a measure of the arc rate of change of **B**. This leads to the definition

$$\tau = \left|\frac{d\mathbf{B}}{ds}\right|. \tag{3.23}$$

The larger the value of τ, the "quicker" the curve moves away from the osculating plane at P.

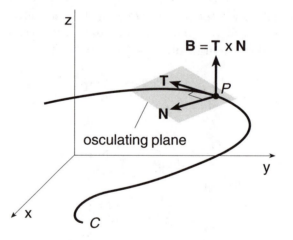

FIGURE 3.6
Geometric view of osculating plane

In fact, it can be shown that the condition for a curve to be planar is that it has zero torsion. Now, because

$$\mathbf{N} = \mathbf{B} \times \mathbf{T} \quad \text{and}$$

$$\frac{d\mathbf{N}}{ds} = \frac{d\mathbf{B}}{ds} \times \mathbf{T} + \mathbf{B} \times \frac{d\mathbf{T}}{ds}$$

$$= -\tau(\mathbf{N} \times \mathbf{T}) + \mathbf{B} \times (\kappa\mathbf{N})$$

$$= -\tau(-\mathbf{B}) - \kappa\mathbf{T}$$

$$= \tau\mathbf{B} - \kappa\mathbf{T}.$$

The three formulae:

$$\frac{d\mathbf{T}}{ds} = \kappa\mathbf{N}$$

$$\frac{d\mathbf{B}}{ds} = -\tau\mathbf{N}$$

$$\frac{d\mathbf{N}}{ds} = \tau\mathbf{B} - \kappa\mathbf{T},$$

are called the **Serret–Frenet** equations and are fundamental to the study of the differential geometry of curves.

REMARK 3.1   If the position vector $\mathbf{R} = x\mathbf{i} + y\mathbf{j} + z\mathbf{k}$ of a particle moving along a curve in space is known, then its velocity and acceleration vectors can easily be determined as follows. We know the velocity vector is

$$\mathbf{V}(t) = \frac{d\mathbf{R}}{dt} = \frac{d\mathbf{R}}{ds}\left(\frac{ds}{dt}\right) = V\frac{d\mathbf{R}}{ds}$$

where V is the speed of the particle and is defined by $V = \frac{ds}{dt}$. Now the acceleration vector $\mathbf{a}(t)$ can be determined in the following manner:

$$\mathbf{a}(t) = \frac{d\mathbf{V}}{dt} = \frac{d^2\mathbf{R}}{dt^2}$$

$$= \mathbf{T}\frac{dV}{dt} + V\frac{d\mathbf{T}}{dt}$$

$$= \mathbf{T}\frac{dV}{dt} + V\left(\frac{d\mathbf{T}}{ds}\right)\left(\frac{ds}{dt}\right)$$

$$= \mathbf{T}\frac{dV}{dt} + V^2\left(\frac{d\mathbf{T}}{ds}\right)$$

$$= \mathbf{T}\frac{dV}{dt} + \kappa V^2\mathbf{N}.$$

Because $\mathbf{T} \cdot \mathbf{T} = 1$, it follows that $\mathbf{T} \cdot \frac{d\mathbf{T}}{ds} = 0$, which implies that $\frac{d\mathbf{T}}{ds}$ is a vector normal to $\mathbf{T}$ and consequently $\frac{d\mathbf{T}}{ds} \propto \mathbf{N}$ where $\mathbf{N}$ is a unit normal vector to $\mathbf{T}$. Thus, $\frac{d\mathbf{T}}{ds} = \kappa\mathbf{N}$, where κ is a constant of proportionality, usually known as the curvature of the curve C. Hence, we have that

$$\left|\frac{d\mathbf{T}}{ds}\right| = \kappa.$$

Let $a_T = \frac{dV}{dt}$ and $a_N = \kappa V^2$. Then the acceleration vector will be given by

$$\mathbf{a}(t) = a_T\mathbf{T} + a_N\mathbf{N}$$
$$|\mathbf{a}(t)| = a_T^2 + a_N^2.$$

To find the formula for the tangential and normal acceleration components, we start by taking the dot product

$$\begin{aligned}\mathbf{V}\cdot\mathbf{a} &= \mathbf{V}\cdot(a_T\mathbf{T} + a_N\mathbf{N})\\ &= V\mathbf{T}\cdot(a_T\mathbf{T} + a_N\mathbf{N})\\ &= Va_T, \text{ so that finally we have}\\ a_T &= \frac{\mathbf{V}\cdot\mathbf{a}}{V} = \frac{\dot{\mathbf{R}}\cdot\ddot{\mathbf{R}}}{|\dot{\mathbf{R}}|}\\ &= \text{Tangential acceleration.}\end{aligned}$$

Alternatively, if we take the cross product, we will obtain

$$\begin{aligned}\mathbf{V}\times\mathbf{a} &= V\mathbf{T}\times(a_T\mathbf{T} + a_N\mathbf{N})\\ &= Va_N(\mathbf{T}\times\mathbf{N})\\ &= Va_N\mathbf{B}, \text{ from which we get}\\ a_N &= \frac{|\mathbf{V}\times\mathbf{a}|}{V} = \frac{|\dot{\mathbf{R}}\times\ddot{\mathbf{R}}|}{|\dot{\mathbf{R}}|}\\ &= \text{Normal acceleration.}\end{aligned}$$

To find the curvature, κ, in terms of the position vector, observe the following

$$\kappa = \left|\frac{d\mathbf{T}}{ds}\right| = \frac{\left|\frac{d\mathbf{T}}{dt}\right|}{\left|\frac{ds}{dt}\right|} = \frac{\left|\frac{d\mathbf{T}}{dt}\right|}{|\dot{\mathbf{R}}|}.$$

We know that $\mathbf{V} = \frac{d\mathbf{R}}{dt} = V\mathbf{T}$ and that

$$\mathbf{a} = \mathbf{T}\frac{dV}{dt} + V\left|\frac{d\mathbf{T}}{dt}\right|\mathbf{N}.$$

We have seen that the vector product $\mathbf{V}\times\mathbf{a}$ can be written as

$$|\mathbf{V} \times \mathbf{a}| = V^2 \left| \frac{d\mathbf{T}}{dt} \right|$$

from which we have

$$\left| \frac{d\mathbf{T}}{dt} \right| = \frac{|\mathbf{V} \times \mathbf{a}|}{V^2} \quad \text{and since}$$

$$\kappa = \frac{\left| \frac{d\mathbf{T}}{dt} \right|}{V} \quad \text{then we have}$$

$$= \frac{|\dot{\mathbf{R}} \times \ddot{\mathbf{R}}|}{|\dot{\mathbf{R}}|^3} \quad \text{in terms of the position vector.}$$

The last line follows from the fact that $\mathbf{V} = \dot{\mathbf{R}}$, $\mathbf{a} = \ddot{\mathbf{R}}$ and $V = |\dot{\mathbf{R}}|$.

$$\frac{d}{ds}(\mathbf{T} \cdot \mathbf{B}) = \frac{d\mathbf{T}}{ds} \cdot \mathbf{B} + \mathbf{T} \cdot \frac{d\mathbf{B}}{ds}$$

$$= \kappa \mathbf{N} \cdot \mathbf{B} + \mathbf{T} \cdot \frac{d\mathbf{B}}{ds}$$

$$= \mathbf{T} \cdot \frac{d\mathbf{B}}{ds}$$

$$\frac{d}{ds}(\mathbf{T} \cdot \mathbf{B}) = 0 = \mathbf{T} \cdot \frac{d\mathbf{B}}{ds}.$$

Thus, $\mathbf{T} \cdot \frac{d\mathbf{B}}{ds} = 0$, which implies that $\frac{d\mathbf{B}}{ds}$ is perpendicular to $\mathbf{T}$. Also, because $\mathbf{B}$ is a unit vector, $\frac{d\mathbf{B}}{ds}$ is perpendicular to $\mathbf{B}$. Therefore, $\frac{d\mathbf{B}}{ds}$ must be parallel to $\mathbf{N}$. Therefore, we can define

$$\frac{d\mathbf{B}}{ds} = -\tau \mathbf{N}$$

$$\left| \frac{d\mathbf{B}}{ds} \right| = \tau$$

where τ is the torsion of the curve C which measures the rate at which the osculating plane turns as we move along C. ∎

Example 3.3
Determine the unit vectors $\mathbf{T}$, $\mathbf{N}$, $\mathbf{B}$ and the constants $\kappa =$ curvature and $\tau =$ torsion for the helix

$$\mathbf{R} = \mathbf{R}(t) = \mathbf{i} \cos t + \mathbf{j} \sin t + \mathbf{k} t.$$

SOLUTION

$$\mathbf{T} = \frac{d\mathbf{R}}{ds} = \frac{d\mathbf{R}}{dt}\left(\frac{dt}{ds}\right) = \frac{\frac{d\mathbf{R}}{dt}}{\left(\frac{ds}{dt}\right)} = \frac{\frac{d\mathbf{R}}{dt}}{\left|\frac{d\mathbf{R}}{dt}\right|}$$

so that we have the following

$$\frac{d\mathbf{R}}{dt} = -\mathbf{i}\sin t + \mathbf{j}\cos t + \mathbf{k}$$

$$\left|\frac{d\mathbf{R}}{dt}\right| = \sqrt{\sin^2 t + \cos^2 t + 1} = \sqrt{2} = \left(\frac{ds}{dt}\right).$$

Thus $\mathbf{T} = \frac{-\mathbf{i}\sin t + \mathbf{j}\cos t + \mathbf{k}}{\sqrt{2}}$. Now we have

$$\frac{d\mathbf{T}}{dt} = \frac{1}{\sqrt{2}}(-\mathbf{i}\cos t - \mathbf{j}\sin t)$$

$$\frac{d\mathbf{T}}{ds} = \frac{d\mathbf{T}}{dt}\left(\frac{dt}{ds}\right) = \frac{\frac{d\mathbf{T}}{dt}}{\left(\frac{ds}{dt}\right)}$$

$$= \frac{1}{\sqrt{2}}\left(\frac{d\mathbf{T}}{dt}\right) = \frac{1}{2}(-\mathbf{i}\cos t - \mathbf{j}\sin t)$$

$$\frac{d\mathbf{T}}{ds} = \kappa\mathbf{N}, \quad \kappa = \left|\frac{d\mathbf{T}}{ds}\right| = \frac{1}{2}$$

and $\mathbf{N} = \frac{d\mathbf{T}/ds}{\kappa} = \frac{\frac{1}{2}(-\mathbf{i}\cos t - \mathbf{j}\sin t)}{\frac{1}{2}} = -\mathbf{i}\cos t - \mathbf{j}\sin t$. Using this quantity, we can then find

$$\mathbf{B} = \mathbf{T}\times\mathbf{N} = \frac{1}{\sqrt{2}}\begin{vmatrix} \mathbf{i} & \mathbf{j} & \mathbf{k} \\ -\sin t & \cos t & 1 \\ -\cos t & -\sin t & 0 \end{vmatrix}$$

$$= \frac{1}{\sqrt{2}}(\mathbf{i}\sin t - \mathbf{j}\cos t + \mathbf{k}).$$

Therefore,

$$\frac{d\mathbf{B}}{ds} = \frac{\frac{d\mathbf{B}}{dt}}{\left(\frac{ds}{dt}\right)} = \frac{\frac{1}{\sqrt{2}}(\mathbf{i}\cos t + \mathbf{j}\sin t)}{\sqrt{2}}$$

$$= \frac{1}{2}(\mathbf{i}\cos t + \mathbf{j}\sin t)$$

$$\left|\frac{d\mathbf{B}}{ds}\right| = \frac{1}{2}(1) = \frac{1}{2} = \tau = \quad \text{torsion.}$$

∎

Example 3.4

Show that for a smooth curve $y = y(x)$ in the xy-plane

$$\kappa(x) = \frac{|y''|}{[1 + (y')^2]^{3/2}}.$$

▯

SOLUTION   The parametric equation to the curve is given by $x = x$ and $y = y(x)$. Then our position, velocity and acceleration vectors, respectively, are given by

$$\mathbf{R} = x\mathbf{i} + y\mathbf{j} + 0\mathbf{k}$$
$$\dot{\mathbf{R}} = \mathbf{i} + y'\mathbf{j} + 0\mathbf{k}$$
$$\ddot{\mathbf{R}} = 0\mathbf{i} + y''\mathbf{j} + 0\mathbf{k}.$$

Now we know that the curvature, κ, is given by

$$\kappa = \frac{|\dot{\mathbf{R}} \times \ddot{\mathbf{R}}|}{|\dot{\mathbf{R}}|^3} \quad \text{which then becomes}$$

$$= \frac{|y''|}{(1 + (y')^2)^{3/2}}.$$

The last line follows because the quantity in the numerator is

$$\dot{\mathbf{R}} \times \ddot{\mathbf{R}} = \begin{vmatrix} \mathbf{i} & \mathbf{j} & \mathbf{k} \\ 1 & y' & 0 \\ 0 & y'' & 0 \end{vmatrix} = \mathbf{k}y''.$$

∎

REMARK 3.2   We note here that both the curvature and torsion are constant for every point on the curve. It can be shown that the circular helix is the only curve for which this is true.   ∎

Although brief mention has been made of the osculating plane, we have not yet discussed how such a plane can be obtained. A close examination of Figure 3.6 will reveal that the vectors associated with the plane all appear to be in the same plane. If we let $P(x, y, z)$ be a general point in the osculating plane and If we let $P_1(x_1, y_1, z_1)$ be another vector in the plane, then we can impose the condition that the vector joining P to P_1 be coplanar with the vectors $\mathbf{T}$ and $d\mathbf{T}/ds$ at P_1. It follows that we can apply the scalar triple product to these three vectors and equate it to zero to obtain the equation of the required plane. In other words, we have

$$\mathbf{P_1P} \cdot \mathbf{T} \times \frac{d\mathbf{T}}{ds} = 0.$$

Example 3.5
Find the equation of the osculating plane to the curve $x = \cos t$, $y = \sin t$, $z = t$ at the point $P_1(1, 0, 0)$ when $t = 0$.   ⬚

SOLUTION   We know that $\mathbf{R} = \mathbf{i}\cos t + \mathbf{j}\sin t + \mathbf{k}t$, $\left|\frac{d\mathbf{R}}{dt}\right| = \sqrt{2}$ and

$$\mathbf{T} = \frac{1}{2}(-\mathbf{i}\sin t + \mathbf{j}\cos t + \mathbf{k})$$
$$\frac{d\mathbf{T}}{ds} = \frac{1}{2}(-\mathbf{i}\cos t - \mathbf{j}\sin t)$$
$$\mathbf{T}(0) = \frac{1}{\sqrt{2}}(-\mathbf{i}(0) + \mathbf{j}(1) + \mathbf{k})$$
$$\frac{d\mathbf{T}}{ds}(0) = -\frac{1}{2}(\mathbf{i}(1) + \mathbf{j}(0))$$
$$\mathbf{P_1P} = \mathbf{i}(x - 1) + \mathbf{j}(y - 0) + \mathbf{k}(z - 0).$$

Thus, the equation of the osculating plane is given by

$$\mathbf{P_1P} \cdot \mathbf{T} \times \frac{d\mathbf{T}}{ds} = 0$$

$$\begin{vmatrix} x & y & z \\ 0 & \frac{1}{\sqrt{2}} & \frac{1}{\sqrt{2}} \\ -\frac{1}{2} & 0 & 0 \end{vmatrix} = 0,$$

i.e. $\quad y\left(-\frac{1}{2\sqrt{2}}\right) + z\left(\frac{1}{2\sqrt{2}}\right) = 0, \quad y = z.$

∎

There is an easier alternate formula to determine the osculating plane. Here we define $\mathbf{T}$ in the following manner.

$$\mathbf{T} = \frac{d\mathbf{R}}{ds} = \frac{d\mathbf{R}}{dt}\left(\frac{dt}{ds}\right)$$

$$\frac{d\mathbf{T}}{ds} = \frac{d^2\mathbf{R}}{dt^2}\left(\frac{dt}{ds}\right)^2 + \frac{d\mathbf{R}}{dt}\left(\frac{d^2t}{ds^2}\right).$$

We can now define the scalar triple product as follows

$$\mathbf{P_1P} \cdot \mathbf{T} \times \frac{d\mathbf{T}}{ds} = 0$$

$$\mathbf{P_1P} \cdot \left(\frac{d\mathbf{R}}{dt}\right)\left(\frac{dt}{ds}\right) \times \left[\frac{d^2\mathbf{R}}{dt^2}\left(\frac{dt}{ds}\right)^2 + \frac{d\mathbf{R}}{dt}\left(\frac{d^2t}{ds^2}\right)\right] = 0$$

$$\mathbf{P_1P} \cdot \left[\frac{d\mathbf{R}}{dt} \times \frac{d^2\mathbf{R}}{dt^2}\right]\left(\frac{dt}{ds}\right)^3 = 0$$

$$\mathbf{P_1P} \cdot \left\{\frac{d\mathbf{R}}{dt} \times \frac{d^2\mathbf{R}}{dt^2}\right\}_{t_0} = 0.$$

This final equation can be rewritten with the notation

$$\mathbf{P_1P} \cdot \dot{\mathbf{R}} \times \ddot{\mathbf{R}} = 0. \tag{3.24}$$

The following example illustrates the use of this simple formula.

Example 3.6
What is the equation of the osculating plane to the space curve $x = t^4$, $y = t^2$, $z = t^3$ at the point $P_1(1, 1, 1)$ when $t = 1$? ☐

SOLUTION   Let our position vector $\mathbf{R}$ be given by $\mathbf{R}(t) = t^4\mathbf{i} + t^2\mathbf{j} + t^3\mathbf{k}$. Then our corresponding derivatives are given by

$$\dot{\mathbf{R}} = (4t^3, 2t, 3t^2)$$
$$\ddot{\mathbf{R}} = (12t^2, 2, 6t) \quad \text{and our vector is given by}$$
$$\mathbf{P_1P} = (x - 1, y - 1, z - 1)$$

at $t = 1$ the derivatives above become $\dot{\mathbf{R}} = (4, 2, 3)$ and $\ddot{\mathbf{R}} = (12, 2, 6)$. Then, forming the triple product, we have

$$P_1 P \cdot \dot{\mathbf{R}} \times \ddot{\mathbf{R}} = \begin{vmatrix} x-1 & y-1 & z-1 \\ 4 & 2 & 3 \\ 12 & 2 & 6 \end{vmatrix} = 0$$

$$3(x-1) + 6(y-1) - 8(z-1) = 0$$
$$3x + 6y - 8z = 1$$

which is the equation of the osculating plane.   ∎

Summary of Formulae

The following are important formulae from the preceding chapter. Suppose $\mathbf{R} = \mathbf{R}(t)$ is a position vector as discussed in the chapter. Then we have:

1. $\mathbf{R} = x\mathbf{i} + y\mathbf{j} + z\mathbf{k}$.

2. $\mathbf{T} = \dfrac{\frac{d\mathbf{R}}{dt}}{\left|\frac{d\mathbf{R}}{dt}\right|} = \dfrac{d\mathbf{R}}{ds}$.

3. $\mathbf{N} = \dfrac{\frac{d\mathbf{T}}{dt}}{\left|\frac{d\mathbf{T}}{dt}\right|}$ unit normal.

4. $\dfrac{d\mathbf{T}}{ds} = \kappa\mathbf{N}$.

5. $\kappa = \left|\dfrac{d\mathbf{T}}{ds}\right| = \dfrac{\left|\frac{d\mathbf{T}}{dt}\right|}{\left(\frac{ds}{dt}\right)} = $ curvature.

6. $\mathbf{B} = \mathbf{T} \times \mathbf{N}$ unit binormal.

7. $\dfrac{d\mathbf{B}}{ds} = \mathbf{T} \times \dfrac{d\mathbf{N}}{ds} = -\tau\mathbf{N}$.

8. $\tau = \left|\dfrac{d\mathbf{B}}{ds}\right| = $ torsion.

9. $\dfrac{d\mathbf{N}}{ds} = \tau\mathbf{B} - \kappa\mathbf{T}$.

10. Serret–Frenet Formulae:

   (a) $\dfrac{d\mathbf{T}}{ds} = \kappa\mathbf{N}$

   (b) $\dfrac{d\mathbf{B}}{ds} = -\tau\mathbf{N}$

   (c) $\dfrac{d\mathbf{N}}{ds} = \tau\mathbf{B} - \kappa\mathbf{T}$.

Exercises

1. Let $\mathbf{R}(t) = \sin t\,\mathbf{i} + \cos t\,\mathbf{j} + \mathbf{k}$.

   (a) Find $\mathbf{R}'(t)$
   (b) Show that $\mathbf{R}'(t)$ is always parallel to the xy-plane
   (c) For what values of t is $\mathbf{R}'(t)$ parallel to the xz-plane?
   (d) Does $\mathbf{R}(t)$ have constant magnitude?
   (e) Does $\mathbf{R}'(t)$ have constant magnitude?
   (f) Compute $\mathbf{R}''(t)$.

2. Find $\mathbf{R}'(t)$ in each of the following cases:

   (a) $\mathbf{R}(t) = \sin t\,\mathbf{i} + e^{-t}\mathbf{j} + 3\mathbf{k}$
   (b) $\mathbf{R}(t) = (e^t\mathbf{i} + \mathbf{j} + t^2\mathbf{k}) \times (t^3\mathbf{i} + \mathbf{j} - \mathbf{k})$
   (c) $\mathbf{R}(t) = (\sin t + t^3)(\mathbf{i} + \mathbf{j} + 2\mathbf{k})$.

3. Show that
$$\frac{d}{dt}\left(\mathbf{R} \times \frac{d\mathbf{R}}{dt}\right) = \mathbf{R} \times \frac{d^2\mathbf{R}}{dt^2}.$$

4. For the curve $x = \sin t - t\cos t$, $y = \cos t + t\sin t$, $z = t^2$, find

   (a) the arc length between $(0,1,0)$ and $(-2\pi,\, 1,\, 4\pi^2)$
   (b) $\mathbf{T}(t)$
   (c) $\mathbf{T}(\pi)$.

In the following four problems, the coordinates of a moving particle are given as a function of time t. Find (a) the speed, (b) the tangential and normal components of acceleration, (c) the unit tangent vector $\mathbf{T}$, and (d) the curvature of the curve, as functions of time.

5. $x = e^t\cos t, y = e^t\sin t, z = 0$.

6. $x = 3t\cos t, y = 3t\sin t, z = 4t$.

7. $x = e^t\cos t, y = e^t\sin t, z = e^t$.

8. $x = 5\sin 4t, y = 5\cos 4t, z = 10t$.

9. The position vector of a moving particle is
$$\mathbf{R}(t) = \cos t(\mathbf{i} - \mathbf{j}) + \sin t(\mathbf{i} + \mathbf{j}) + \frac{1}{2}t\mathbf{k}$$

   (a) Determine the velocity and speed of the particle.

(b) Determine the acceleration of the particle.

(c) Find a unit tangent to the path of the particle, in the direction of motion.

(d) Show that the curve traversed by the particle has constant curvature κ, and find its value.

For the following problems, find (a) the unit tangent and unit normal vectors $\mathbf{T}(t)$ and $\mathbf{N}(t)$, (b) the curvature.

10. $\mathbf{R}(t) = (\sin 4t, 3t, \cos 4t)$.

11. $\mathbf{R}(t) = (6t, 3\sqrt{2}t^2, 2t^3)$.

12. $\mathbf{R}(t) = (\sqrt{2}\cos t, \sin t, \sin t)$.

13. $\mathbf{R}(t) = (t^2, 2t, \ln t)$.

14. Find the vectors $\mathbf{T}$, $\mathbf{N}$, and $\mathbf{B}$ at the given point

   (a) $\mathbf{R}(t) = (t^2, \frac{2}{3}t^3, t)$, $(1, \frac{2}{3}, 1)$
   (b) $\mathbf{R}(t) = (e^t, e^t \sin t, e^t \cos t)$, $(1, 0, 1)$.

15. Find the equations of the normal plane and osculating plane of the curve at the given point

   (a) $x = 2\sin 3t, y = t, z = 2\cos 3t$, $(0, \pi, -2)$
   (b) $x = t, y = t^2, z = t^3$, $(1, 1, 1)$.

16. The position function of a particle is given by $\mathbf{R}(t) = (t^2, 5t, t^2 - 16t)$. When is the speed a minimum?

17. A force with magnitude 20N acts directly upward from the xy-plane on an object with mass 4kg. The object starts at the origin with initial velocity $\mathbf{V}(0) = \mathbf{i} - \mathbf{j}$. Find its position function and its speed at time t.

18. The position function of a spaceship is

$$\mathbf{R}(t) = (3 + t)\mathbf{i} + (2 + \ln t)\mathbf{j} + (z - \frac{4}{t^2 + 1})\mathbf{k}$$

and the coordinates of a space station are (6,4,9). If the captain wants the spaceship to coast into the space station, when should the engines be turned off?

19. Let C be the curve with equations $x = 2 - t^3$, $y = 2t - 1$, $z = \ln t$. Find

(a) the point where C intersects the xz-plane

(b) parametric equations of the tangent line (1,1,0) and

(c) an equation of the normal plane to C at (1,1,0).

20. (a) Find an equation of the osculating plane of the curve $x = \sin 2t$, $y = t$, $z = \cos 2t$ at the point $(0, \pi, 1)$.

(b) A particle starts at the origin with initial velocity $\mathbf{i} + 2\mathbf{j} + \mathbf{k}$. Its acceleration is $\mathbf{a}(t) = t\mathbf{i} + \mathbf{j} + t^2\mathbf{k}$. Find its position function.

(c) Find the tangential and normal components of the acceleration vector of a particle with position function $\mathbf{R}(t) = t\mathbf{i} + 2t\mathbf{j} + t^2\mathbf{k}$.

21. If $\mathbf{P} = \mathbf{A}\cos\omega t + \mathbf{B}\sin\omega t$, where $\mathbf{A}$ and $\mathbf{B}$ are arbitrary constant vectors, show that $\mathbf{P} \times \frac{d\mathbf{P}}{dt}$ is a constant and that $\frac{d^2\mathbf{P}}{dt^2} + \omega^2\mathbf{P} = \mathbf{0}$.

22. If $\mathbf{R} = t^2\mathbf{i} - t^3\mathbf{j} + t^4\mathbf{k}$ is the vector from the origin to a moving particle, find the velocity of the particle when $t = 1$. What is the component of this velocity in the direction of the vector $8\mathbf{i} - \mathbf{j} + 4\mathbf{k}$? What is the vector acceleration of the particle? What are the tangential and normal components of this acceleration?

23. If a particle starts to move from rest at the point (0,1,2) with component acceleration $a_x = 1 + t$, $a_y = t^3$, $a_z = 2t - t^2$ find the vector from the origin to the instantaneous position of the particle.

24. At $t = 0$ a body starts to move from the point $(-1, 0, 1)$ with velocity $v_0 = \mathbf{i} + 2\mathbf{j} - 3\mathbf{k}$. If the acceleration of the body is $a(t) = 3t^2\mathbf{i} + \mathbf{j} + (\cos 2t)\mathbf{k}$, find the position of the body at any subsequent time. At what velocity is the body moving at $t = 1$?

Chapter 4

The Del Operator

Let $\phi(x, y, z)$ be a scalar function of position possessing first partial derivatives with respect to x, y, and z throughout some region of space, and let $\mathbf{R} = x\mathbf{i} + y\mathbf{j} + z\mathbf{k}$ be the vector drawn from the origin to a general point P : (x, y, z). If we move from P to a neighboring point Q : $(x+\Delta x, y+\Delta y, z+\Delta z)$ (Figure 4.1), the function ϕ will change by an amount $\Delta\phi$ whose exact value, as derived in calculus, is

$$\Delta\phi = \phi(x + \Delta x, y + \Delta y, z + \Delta z) - \phi(x, y, z).$$

By Taylor series expansion, the first term on the right hand side can be written as

$$\phi(x + \Delta x, y + \Delta y, z + \Delta z) = \phi(x, y, z) + \frac{\partial\phi}{\partial x}\Delta x + \frac{\partial\phi}{\partial y}\Delta y + \frac{\partial\phi}{\partial z}\Delta z$$
$$+ \mathrm{O}\left(\Delta x^2, \Delta y^2, \Delta z^2\right)$$

where O stands for order of expansion and contains all the remaining terms of the Taylor series expansion. Therefore

$$\Delta\phi = \frac{\partial\phi}{\partial x}\Delta x + \frac{\partial\phi}{\partial y}\Delta y + \frac{\partial\phi}{\partial z}\Delta z$$
$$+ \mathrm{O}(\Delta x^2, \Delta y^2, \Delta z^2). \tag{4.1}$$

Now, if we divide the change $\Delta\phi$ by the distance $\Delta s = |\Delta\mathbf{R}|$ between P and Q, we obtain a measure of the rate at which ϕ changes when we move from P to Q:

$$\frac{\Delta\phi}{\Delta s} = \frac{\partial\phi}{\partial x}\frac{\Delta x}{\Delta s} + \frac{\partial\phi}{\partial y}\frac{\Delta y}{\Delta s} + \frac{\partial\phi}{\partial z}\frac{\Delta z}{\Delta s}$$
$$+ \mathrm{O}\left(\left(\frac{\partial x}{\partial s}\right)\Delta x, \left(\frac{\partial y}{\partial s}\right)\Delta y, \left(\frac{\partial z}{\partial s}\right)\Delta z\right) \tag{4.2}$$

where the limiting value of the last term on the right-hand side goes to zero
as Δx, Δy, and Δz go to zero.

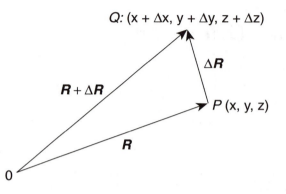

FIGURE 4.1
Coordinate vectors of two neighbouring points

For example, if $\phi(x, y, z)$ is the temperature at the general point $P(x, y, z)$,
then $\Delta\phi/\Delta s$ is the average rate of change of temperature in the direction
at which Δs is measured. The limiting value of $\Delta\phi/\Delta s$ as Q approaches P
along the segment PQ is called the **derivative** of ϕ **in the direction of**
PQ or simply the **directional derivative** of ϕ. Clearly, in the limit the
last three terms in (4.2) become zero and we have explicitly

$$\frac{d\phi}{ds} = \frac{\partial\phi}{\partial x}\frac{dx}{ds} + \frac{\partial\phi}{\partial y}\frac{dy}{ds} + \frac{\partial\phi}{\partial z}\frac{dz}{ds}. \tag{4.3}$$

The first factor in each product on the right in (4.3) depends only on ϕ
and the coordinates of the point at which the derivatives of ϕ are evaluated.
The second factor in each product is independent of ϕ and only depends on
the direction in which the derivative is being computed. This observation
suggests that $\frac{d\phi}{ds}$ can be thought of as the dot product of two vectors, one
depending only on ϕ and the coordinates of P, the other depending only in
the direction of ds. In fact, we can write

$$\frac{d\phi}{ds} = \left(\frac{\partial\phi}{\partial y}\mathbf{i} + \frac{\partial\phi}{\partial y}\mathbf{j} + \frac{\partial\phi}{\partial z}\mathbf{k}\right) \cdot \left(\frac{dx}{ds}\mathbf{i} + \frac{dy}{ds}\mathbf{j} + \frac{dz}{ds}\mathbf{k}\right)$$

$$= \left(\frac{\partial\phi}{\partial x}\mathbf{i} + \frac{\partial\phi}{\partial y}\mathbf{j} + \frac{\partial\phi}{\partial z}\mathbf{k}\right) \cdot \frac{d\mathbf{R}}{ds} \tag{4.4}$$

we have thus introduced the vector function,

$$\frac{\partial\phi}{\partial x}\mathbf{i} + \frac{\partial\phi}{\partial y}\mathbf{j} + \frac{\partial\phi}{\partial z}\mathbf{k}$$

which is known as the **gradient of** ϕ or simply **grad** ϕ, and in this notation equation (4.4) can be rewritten as

$$\frac{d\phi}{ds} = (\mathbf{grad}\ \phi) \cdot \frac{d\mathbf{R}}{ds}. \tag{4.5}$$

To determine the significance of **grad** ϕ, we observe first that since Δs is by definition just the length of $\Delta \mathbf{R}$, it follows that $\frac{d\mathbf{R}}{ds}$ is a unit vector. Hence, the dot product $(\mathbf{grad}\ \phi) \cdot \frac{d\mathbf{R}}{ds}$ is just the projection of **grad** ϕ in the direction of $\frac{d\mathbf{R}}{ds}$. Furthermore, **grad** ϕ has the property that its projection in any direction is equal to the derivative of ϕ in that direction (see Figure 4.2).

Because the maximum projection of a vector is the vector itself, it is clear that **grad** ϕ extends in the direction of the greatest rate of change of ϕ and has that rate of change for its length.

If we set $\phi(x, y, z) = c$, we obtain, as c takes on different values, a family of surfaces known as **level surfaces** of ϕ. And, on the assumption that ϕ is single-valued, one and only one level surface that passes through P, and fix our attention on neighboring points Q that lie on this surface, we have $\frac{d\phi}{ds} = 0$, because $\Delta\phi = 0$. Because, by definition, ϕ has the same value at all points of a level surface. Therefore, by (4.5)

$$(\mathbf{grad}\ \phi) \cdot \frac{d\mathbf{R}}{ds} = 0 \tag{4.6}$$

for any vector $\frac{d\mathbf{R}}{ds}$ that has the limiting direction of a secant PQ of the level surface. Clearly, such vectors are all tangent to $\phi = c$ at the point P. Hence, from the vanishing of the dot product in (4.6) it follows that **grad** ϕ is perpendicular to every tangent to the level surface at P. In other words, the gradient of ϕ at any point is perpendicular to the level surface of ϕ that passes through that point (see Figure 4.2 (b)).

Evidently, **grad** ϕ is related to the level surfaces of ϕ in a way that is independent of the particular coordinate system used to describe ϕ. In other words, **grad** ϕ depends only on the intrinsic properties of ϕ. Therefore, in the expression

$$\mathbf{grad}\ \phi = \frac{\partial\phi}{\partial x}\mathbf{i} + \frac{\partial\phi}{\partial y}\mathbf{j} + \frac{\partial\phi}{\partial z}\mathbf{k} \tag{4.7}$$

i, **j**, and **k** can be replaced by any other set of mutually perpendicular unit vectors provided that $\frac{\partial\phi}{\partial x}$, $\frac{\partial\phi}{\partial y}$, $\frac{\partial\phi}{\partial z}$ are replaced by the directional derivatives of ϕ along the new axes. The gradient of a function is frequently written in its operational form as

$$\mathbf{grad}\ \phi = \left(\mathbf{i}\frac{\partial}{\partial x} + \mathbf{j}\frac{\partial}{\partial y} + \mathbf{k}\frac{\partial}{\partial z}\right)\phi.$$

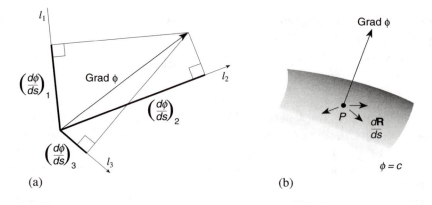

FIGURE 4.2
Geometric interpretation of the gradient

The operational *vector* or differential operator is denoted by the symbol ∇ (read *del*), i.e.,

$$\nabla \equiv \mathbf{i}\frac{\partial}{\partial x} + \mathbf{j}\frac{\partial}{\partial y} + \mathbf{k}\frac{\partial}{\partial z}$$

which leads to the notation

$$\mathbf{grad}\ \phi = \nabla \phi$$
$$\frac{d\phi}{ds} = \nabla \phi \cdot \frac{d\mathbf{R}}{ds}.$$

In addition, if ϕ is a function of a single variable u, which in turn is a function of x, y, and z, then

$$\begin{aligned}
\nabla \phi &= \frac{\partial \phi}{\partial x}\mathbf{i} + \frac{\partial \phi}{\partial y}\mathbf{j} + \frac{\partial \phi}{\partial z}\mathbf{k} \\
&= \frac{d\phi}{du}\frac{\partial u}{\partial x}\mathbf{i} + \frac{d\phi}{du}\frac{\partial u}{\partial y}\mathbf{j} + \frac{d\phi}{du}\frac{\partial u}{\partial z}\mathbf{k} \\
&= \frac{d\phi}{du}\left(\frac{\partial u}{\partial x}\mathbf{i} + \frac{\partial u}{\partial y}\mathbf{j} + \frac{\partial u}{\partial z}\mathbf{k}\right) \\
&= \frac{d\phi}{du}\cdot \nabla u.
\end{aligned} \qquad (4.8)$$

Example 4.1
What is the directional derivative of the function $\phi(x, y, z) = xy^2 + yz^3$ at the point $(2, -1, 1)$ in the direction of the vector $\mathbf{i} + 2\mathbf{j} + 2\mathbf{k}$?

⬚

SOLUTION   Our first step must be to find the gradient of ϕ at the point $(2, -1, 1)$. This is

$$\nabla\phi = \frac{\partial(xy^2 + yz^3)}{\partial x}\mathbf{i} + \frac{\partial(xy^2 + yz^3)}{\partial y}\mathbf{j} + \frac{\partial(xy^2 + yz^3)}{\partial z}\mathbf{k} \quad \text{at} \quad (2, -1, 1)$$

$$= y^2\mathbf{i} + (2xy + z^3)\mathbf{j} + 3yz^2\mathbf{k} \quad \text{at} \quad (2, -1, 1)$$

$$= \mathbf{i} - 3\mathbf{j} - 3\mathbf{k}.$$

The projection of this in the direction of the given vector will be the required directional derivative. Because this projection can be found at once as the dot product of $\nabla\phi$ and a unit vector in the given direction, we next reduce $\mathbf{i} + 2\mathbf{j} + 2\mathbf{k}$ to a unit vector by dividing its magnitude as follows:

$$\frac{\mathbf{i} + 2\mathbf{j} + 2\mathbf{k}}{\sqrt{1 + 4 + 4}} = \frac{1}{3}\mathbf{i} + \frac{2}{3}\mathbf{j} + \frac{2}{3}\mathbf{k}.$$

The answer to our problem is now given by

$$\nabla\phi \cdot \left(\frac{1}{3}\mathbf{i} + \frac{2}{3}\mathbf{j} + \frac{2}{3}\mathbf{k}\right) = (\mathbf{i} - 3\mathbf{j} - 3\mathbf{k}) \cdot \left(\frac{1}{3}\mathbf{i} + \frac{2}{3}\mathbf{j} + \frac{2}{3}\mathbf{k}\right)$$

$$= -\frac{11}{3}.$$

The negative sign, of course, indicates that ϕ decreases in the given direction.
∎

DEFINITION 4.1   *If $\phi = \phi(x, y, z)$ is a differentiable function of x,y, and z and $\mathbf{U}$ is any unit vector, then*

$$D_{\mathbf{U}}\phi = \nabla\phi \cdot \mathbf{U},$$

is defined as the directional derivative of $\phi = \phi(x, y, z)$ in the direction of the unit vector $\mathbf{U}$.

Note that it is similar to the formula $\frac{d\phi}{ds} = \nabla\phi \cdot \frac{d\mathbf{R}}{ds}$, provided we substitute $D_{\mathbf{U}}\phi$ for $\frac{d\phi}{ds}$ and $\mathbf{U}$ for $\frac{d\mathbf{R}}{ds}$. Thus, $D_{\mathbf{U}}\phi = \nabla\phi \cdot \mathbf{U}$ shows that the directional derivative of ϕ in the direction of $\mathbf{U}$ is just the scalar projection of $\nabla\phi$ along the direction of $\mathbf{U}$.

Example 4.2
What is the unit normal to the surface $xy^3z^2 = 4$ at the point $(-1, -1, 2)$?
⬚

SOLUTION   Let us regard the given surface as a particular level surface
of the function at the point $(-1, -1, 2)$ will be perpendicular to the level
surface through $(-1, -1, 2)$, which is the given surface. When this gradient
has been found, the unit normal can be obtained at once by dividing the
gradient by its magnitude:

$$\nabla\phi = \frac{\partial(xy^3z^2)}{\partial x}\mathbf{i} + \frac{\partial(xy^3z^2)}{\partial y}\mathbf{j} + \frac{\partial(xy^3z^2)}{\partial z}\mathbf{k} \quad \text{at} \quad (-1, -1, 2)$$

$$= -4\mathbf{i} - 12\mathbf{j} + 4\mathbf{k}$$

$$|\nabla\phi| = \sqrt{16 + 144 + 16} = 4\sqrt{11}$$

$$\frac{\nabla\phi}{|\nabla\phi|} = \frac{-4\mathbf{i} - 12\mathbf{j} + 4\mathbf{k}}{4\sqrt{11}} = -\frac{1}{\sqrt{11}}\mathbf{i} - \frac{3}{\sqrt{11}}\mathbf{j} + \frac{1}{\sqrt{11}}\mathbf{k}.$$

It may be necessary to reverse the direction of this result by multiplying it
by -1, depending on which side of the surface we wish the normal to extend.
∎

Example 4.3
Find the unit normal $\mathbf{n}$ to the surface $x^2 + y^2 - z = 0$ at the point $(1, 1, 2)$.
▯

SOLUTION   Let us define our function to be $f(x, y, z) = x^2 + y^2 - z = 0$, so that the surface is $f = 0$. It follows that $\nabla f = (2x, 2y, -1)$
(using vector notation). At the point (1,1,2), $\nabla f = (2, 2, -1)$. This is a
vector normal to the surface. To find the unit normal we must divide by
the magnitude, which in this case is $(2^2 + 2^2 + 1^2)^{1/2} = 3$. We thus have
$\nabla f/|\nabla f| = (2/3, 2/3, -1/3)$. We note here that the unit normal is not
uniquely defined as mentioned at the end of the previous example. The
vector $-\mathbf{n} = (-2/3, -2/3, 1/3)$ is also a unit normal to the surface.   ∎

4.1   Gradient Characterizes Maximum Increase

We will first show that the length and direction of **grad** f are independent
of the particular choice of Cartesian coordinates. By the definition of a scalar
function, the value of f at a point P depends on P but is independent of the
coordinates, and s, the arc length of that ray C, is also independent of the
choice of coordinates. Hence, the directional derivative is also independent
of particular choice of coordinates. Now we have the directional derivative
defined as

$$D_{\mathbf{U}}\phi = \nabla\phi \cdot \mathbf{U} = |\nabla\phi||\mathbf{U}|\cos\gamma = |\nabla\phi|\cos\gamma \quad (|\mathbf{U}| = 1)$$

where γ is the angle between $\nabla\phi$ and $\mathbf{U}$. Because $0 \le \gamma \le \pi$, we have $-1 \le \cos\gamma \le 1$ and consequently

$$-|\nabla\phi| \le D_{\mathbf{U}}\phi \le |\nabla\phi|.$$

In other words, the maximum value of the directional derivative is $|\nabla\phi|$ and it occurs when $\mathbf{U}$ has the same direction as $\nabla\phi$ (when $\cos\gamma = 1$). The minimum value of the directional derivative is $-|\nabla f|$ and it occurs when $\mathbf{U}$ and $\nabla\phi$ have opposite directions (when $\cos\gamma = -1$).

The gradient $\nabla\phi$ is often spoken of as a *gradient vector field*. It must be noted that $\nabla\phi$ assigns a vector to each point in the domain of ϕ. This gradient vector field has important geometric significance. It shows the direction in which ϕ is increasing the fastest and the direction that is orthogonal to the level surfaces (or curves in the plane) of ϕ.

In order to see this, imagine a hill such as the one shown in Figure 4.3. We let h be the height function, clearly a function of two variables. If the level curves of h are drawn, we see that they are just the contours of the hill. One could imagine them as level paths on the hill (see Figure 4.3 (b)).

From the figure, it is obvious that to get to the top of the hill the fastest, one must walk perpendicular to the level contours. This is consistent with the theory in the preceding paragraphs. Thus we can state that the direction of fastest increase (the gradient) is orthogonal to the level curves, which is the direction $\nabla\phi$ in Figure 4.3 (a).

Example 4.4

Find the directional derivative of $\phi(x, y, z) = xy^2 - 4x^2y + z^2$ at $(1, -1, 2)$ in the direction of $6\mathbf{i} + 2\mathbf{j} + 3\mathbf{k}$. What are the directions of maximum and minimum increase at $(1, -1, 2)$?   □

SOLUTION   We have $\frac{\partial\phi}{\partial x} = y^2 - 8xy$, $\frac{\partial\phi}{\partial y} = 2xy - 4x^2$ and $\frac{\partial\phi}{\partial z} = 2z$ so that

$$\nabla\phi(x, y, z) = (y^2 - 8xy)\mathbf{i} + (2xy - 4x^2)\mathbf{j} + 2z\mathbf{k}$$
$$\nabla\phi(1, -1, 2) = 9\mathbf{i} - 6\mathbf{j} + 4\mathbf{k}.$$

Because $|6\mathbf{i} + 2\mathbf{j} + 3\mathbf{k}| = 7$, then $\mathbf{U} = \frac{6}{7}\mathbf{i} + \frac{2}{7}\mathbf{j} + \frac{3}{7}\mathbf{k}$ is a unit vector in the indicated direction. Then we have

$$D_{\mathbf{U}}\phi(1, -1, 2) = \nabla\phi \cdot \mathbf{U} = (9\mathbf{i} - 6\mathbf{j} + 4\mathbf{k}) \cdot \left(\frac{6}{7}\mathbf{i} + \frac{2}{7}\mathbf{j} + \frac{3}{7}\mathbf{k}\right) = \frac{54}{7}.$$

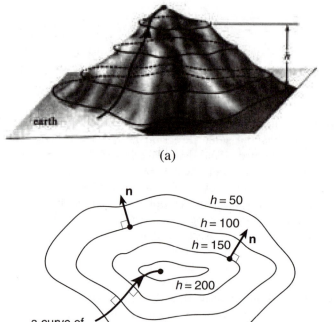

(a)

h = 50
h = 100
h = 150
h = 200

a curve of
steepest ascent
up the hill

(b)

contour map of hill
250 feet high

FIGURE 4.3
Orthogonal contours of a hill

The maximum value of the directional derivative at ϕ at $(1, -1, 2)$ is $|\nabla\phi(1, -1, 2)| = \sqrt{133}$. The minimum value of $D_{\mathbf{U}}\phi(1, -1, 2)$ is then $-\sqrt{133}$. ∎

The previous two statements concerning the maximum and minimum value of the directional derivative can be put another way. The gradient vector $\nabla\phi$ points in the direction in which ϕ increases most rapidly, whereas $-\nabla\phi$ points in the direction of most rapid decrease of ϕ.

Example 4.5
Each year there is a bicycle race in Los Angeles up to the top of a hill by a road known to be the steepest in the city. To understand why a bicyclist will zig-zag up the road, let us suppose the graph of $f(x, y) = 4 - \frac{2}{3}\sqrt{x^2 + y^2}$, $0 \le z \le 4$ shown in Figure 4.3 (a) is a mathematical model of the hill. ⬚

SOLUTION   The gradient of f is

$$\nabla f(x,y) = \frac{2}{3}\left(\frac{-x}{\sqrt{x^2+y^2}}\mathbf{i} + \frac{-y}{\sqrt{x^2+y^2}}\mathbf{j}\right) = \frac{\frac{2}{3}}{\sqrt{x^2+y^2}}\mathbf{R}$$

where $\mathbf{R} = -x\mathbf{i} - y\mathbf{j}$ is a vector pointing to the center of the circular base. Thus, the steepest ascent up the hill is a straight road whose projection in the xy-plane is a radius of the circular base. Obviously, this direction will cause the bicyclist to expend an enormous amount of energy to arrive at the top of the hill. Because of this, a bicyclist will zig-zag, or seek a direction $\mathbf{U}$ other than ∇f, in order to reduce this component. ∎

Example 4.6
The temperature in a rectangular box is approximated by $T(x,y,z) = xyz(1-x)(2-y)(3-z)$, $0 \le x \le 1$, $0 \le y \le 2$, $0 \le z \le 3$. If a mosquito is located at $(\frac{1}{2},1,1)$, in which direction should it fly to cool off as rapidly as possible?

SOLUTION The gradient of T is

$$\nabla T(x,y,z) = yz(2-y)(3-z)(1-2x)\mathbf{i} + xz(1-x)(3-z)(2-2y)\mathbf{j} + xy(1-x)(2-y)(3-2z)\mathbf{k}.$$

Therefore,

$$\nabla T\left(\frac{1}{2},1,1\right) = \frac{1}{4}\mathbf{k}.$$

To cool off most rapidly, the mosquito should fly in the direction of $-\frac{1}{4}\mathbf{k}$, i.e., it should dive for the floor of the box, where the temperature is $T(x,y,z) = 0$. ∎

4.2 Tangent Planes and Normal Lines

Geometric Interpretation of the Gradient — Functions of Two Variables

Suppose $f(x,y) = c$ is the level curve of the differentiable function $z = f(x,y)$ that passes through a specified $P(x_0, y_0)$, i.e., $f(x_0, y_0) = c$. If this level curve is parameterized by the differentiable functions $x = g(t)$, $y = h(t)$ such that

$$x_0 = gt(t_0), \quad y_0 = h(t_0),$$

then the derivative of $f(g(t), h(t)) = c$ with respect to t is

$$\frac{\partial f}{\partial x}\frac{dx}{dt} + \frac{\partial f}{\partial y}\frac{dy}{dt} = 0 \tag{4.9}$$

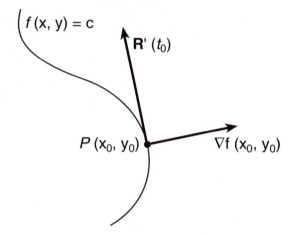

FIGURE 4.4
$\nabla f(x_0, y_0)$ **shown perpendicular to a level curve**

when we introduce the vectors

$$\nabla f(x, y) = \frac{\partial f}{\partial x}\mathbf{i} + \frac{\partial f}{\partial y}\mathbf{j} \quad \text{and} \quad \mathbf{R}'(t) = \frac{dx}{dt}\mathbf{i} + \frac{dy}{dt}\mathbf{j}$$

equation (4.9) becomes $\nabla f \cdot \mathbf{R}'$. Specifically, at $t = t_0$, we have

$$\nabla f(x_0, y_0) \cdot \mathbf{R}'(t_0) = 0. \tag{4.10}$$

Thus, if $\mathbf{R}'(t_0) \neq 0$, the vector $\nabla f(x_0, y_0)$ is orthogonal to the tangent vector $\mathbf{R}'(t_0)$ at $P(x_0, y_0)$. This is interpreted to mean ∇f is perpendicular to the level curve at P (see Figure 4.4)

Example 4.7
Find the level curve of $f(x, y) = -x^2 + y^2$ passing through (2,3). Graph the gradient at the point.

SOLUTION   Because $f(2, 3) = -4 + 9 = 5$, the level curve is the hyperbola $-x^2 + y^2 = 5$. Now

$$\nabla f(x,y) = -2x\mathbf{i} + 2y\mathbf{j} \quad \text{and} \quad \nabla f(2,3) = -5\mathbf{i} + 6\mathbf{j}.$$

Figure 4.5 shows the level curve and $\nabla f(2,3)$.　∎

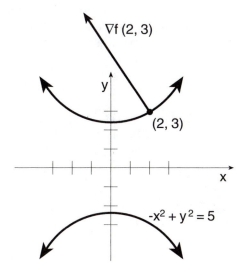

FIGURE 4.5
Level curve and gradient at (2,3)

Geometric Interpretation of the Gradient — Functions of Three Variables

Here again, if we let $F(x,y,z) = c$ be the level surface of a differentiable function $w = F(x,y,z)$ that passes through $P(x_0, y_0, z_0)$. If the differentiable functions

$$x = f(t), \quad y = g(t), \quad z = h(t)$$

are the parametric equations of a curve C on the surface for which $x_0 = f(t_0)$, $y_0 = g(t_0)$, $z_0 = h(t_0)$, then the derivative of $F(f(t), g(t), h(t)) = 0$ implies that

$$\frac{\partial F}{\partial x}\frac{dx}{dt} + \frac{\partial F}{\partial y}\frac{dy}{dt} + \frac{\partial F}{\partial z}\frac{dz}{dt} = 0$$

or $\quad \left(\dfrac{\partial F}{\partial x}\mathbf{i} + \dfrac{\partial F}{\partial y}\mathbf{j} + \dfrac{\partial F}{\partial z}\mathbf{k} \right) \cdot \left(\dfrac{dx}{dt}\mathbf{i} + \dfrac{dy}{dt}\mathbf{j} + \dfrac{dz}{dt}\mathbf{k} \right) = 0. \qquad (4.11)$

In particular, at $t = t_0$ (4.11) becomes

$$\nabla F(x_0, y_0, z_0) \cdot \mathbf{R}'(t_0) = 0 \qquad (4.12)$$

Thus, when $\mathbf{R}'(t_0) \neq 0$, the vector $\nabla F(x_0, y_0, z_0)$ is orthogonal to the tangent vector $\mathbf{R}'(t_0)$. Because this argument holds for any differentiable curve through $P(x_0, y_0, z_0)$ on the surface, we conclude that ∇F is perpendicular (normal) to the level surface at P (see Figure 4.6).

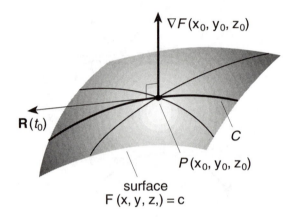

FIGURE 4.6
∇F **shown perpendicular to level surface at** P

Example 4.8
Find the level surface of $F(x, y, z) = x^2 + y^2 + z^2$ passing through $(1, 1, 1)$. Graph the gradient at the point.   ⬚

SOLUTION   Because $F(1, 1, 1) = 3$, the level surface passing through $(1, 1, 1)$ is the sphere $x^2 + y^2 + z^2 = 3$. The gradient of the function is

$$\nabla F(x, y, z) = 2x\mathbf{i} + 2y\mathbf{j} + 2z\mathbf{k}$$

and so, at the given point

$$\nabla F(1, 1, 1) = 2\mathbf{i} + 2\mathbf{j} + 2\mathbf{k}.$$

The level surface and $\nabla F(1, 1, 1)$ are illustrated in Figure 4.7.   ■

4.3   Tangent Plane

In the study of differential calculus a basic problem is finding an equation of a tangent line to the graph of a function. In 3-space the analogous problem is finding an equation of a **tangent plane** to a surface. We assume again that $w = F(x, y, z)$ is a differentiable function and that the surface is given by $F(x, y, z) = c$.

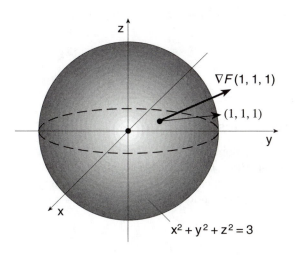

FIGURE 4.7
Level surface and $\nabla F(1, 1, 1)$

DEFINITION 4.2   *Let $P(x_0, y_0, z_0)$ be a point on the graph of $F(x, y, z) = c$ where ∇F is not equal to the zero vector. The tangent plane at P is that plane through P that is perpendicular to ∇F evaluated at P.*

Thus, if $P(x, y, z)$ and $P_1(x_0, y_0, z_0)$ are points on the tangent plane and $\mathbf{R}$ and $\mathbf{R_0}$ are their respective position vectors, then a vector equation of the tangent plane is

$$\nabla F(x_0, y_0, z_0) \cdot (\mathbf{R} - \mathbf{R_0}) = 0. \tag{4.13}$$

Example 4.9
Find an equation of the tangent plane to the graph of $x^2 - 4y^2 + z^2 = 16$ at (2,1,4).   ▯

SOLUTION   By defining $\phi(x, y, z) = x^2 - 4y^2 + z^2$, the given surface is the level surface $\phi(x, y, z) = \phi(2, 1, 4) = 16$ passing through (2,14). Now $\phi_x(x, y, z) = 2x$, $\phi_y(x, y, z) = -8y$, and $\phi_z(x, y, z) = 2z$, so that

$$\nabla\phi(x, y, z) = 2x\mathbf{i} - 8y\mathbf{j} + 2z\mathbf{k} \quad \text{and}$$
$$\nabla\phi(2, 1, 4) = 4\mathbf{i} - 8\mathbf{j} + 8\mathbf{k}.$$

It follows that an equation of the tangent plane is $\nabla\phi \cdot (\mathbf{R} - \mathbf{R}_0) = 0$, i.e.,

$$4(x - 2) - 8(y - 1) + 8(z - 4) = 0 \quad \text{or} \quad x - 2y + 2z = 8.$$

■

4.4   Normal Lines

Let $P(x_0, y_0, z_0)$ be a point on the graph of $\phi(x, y, z) = c$, where $\nabla\phi$ is not $\mathbf{0}$. The line containing $P(x_0, y_0, z_0)$ that is parallel to $\nabla\phi(x_0, y_0, z_0)$ is called the **normal line** to the surface at P. The normal line is perpendicular to the tangent plane to the surface at P. This equation is given by (in parametric form)

$$\mathbf{R} = \mathbf{R}_0 + t(\nabla\phi)_{\mathbf{R}_0}.$$

Example 4.10
Find the parametric equations for the normal line to the surface $\phi = \frac{1}{2}x^2 + \frac{1}{2}y^2 + 4$ at $(1, -1, 5)$.   ⧠

SOLUTION   A normal line equation is given by $\mathbf{R} = \mathbf{R}_0 + t(\nabla\phi)$. A direction vector for the normal line at $(1, -1, 5)$ is $\nabla\phi(1, -1, 5) = \mathbf{i} - \mathbf{j} - \mathbf{k}$. It follows that the parametric equations for the normal line are $x = 1 + t$, $y = -1 - t$, and $z = 5 - t$.   ■

Example 4.11
Find the equations of the tangent plane and the normal line to the surface of the paraboloid of revolution $x^2 + y^2 - z = 1$ at the point $(1, 2, 4)$.   ⧠

SOLUTION   This example will illustrate how to obtain the normal line and tangent plane. First, we can consider $\phi(x, y, z) \equiv x^2 + y^2 - z = d$. Then we have

$$\nabla f = 2x\mathbf{i} + 2y\mathbf{j} - \mathbf{k}$$
$$= 2\mathbf{i} + 4\mathbf{j} - \mathbf{k} \text{ at the given point } (1,2,4).$$

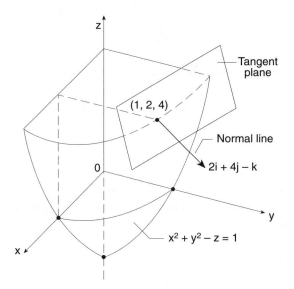

FIGURE 4.8
Paraboloid of revolution $x^2 + y^2 - z = 1$

Therefore, the direction numbers of the normal are simply $(2, 4, -1)$. From this, it follows that the tangent plane has the equation $2x + 4y - z = d$ where d is to be determined. Because we know that the point $(1, 2, 4)$ is in the plane, we can then find the value of d, i.e., $2(1) + 4(2) - 1(4) = d$. Thus the tangent plane is given by $2x + 4y - z = 6$. We can write the equation of the normal line in symmetric form as

$$\mathbf{PP_0} \times \nabla\phi|_{(1,2,4)} = \mathbf{0}$$
$$= \begin{vmatrix} \mathbf{i} & \mathbf{j} & \mathbf{k} \\ x-1 & y-2 & z-4 \\ 2 & 4 & -1 \end{vmatrix}$$
$$= \mathbf{i}[-(y-2) - 4(z-4)] - \mathbf{j}[-(x-1) - 2(z-4)]$$
$$= +\mathbf{k}[4(x-1) - 2(y-2)] = \mathbf{0}.$$

The normal line will then be given by the following equations

$$\frac{(y-2)}{4} = \frac{(z-4)}{-1}$$

$$\frac{(x-1)}{2} = \frac{(z-4)}{-1}$$

$$\frac{(x-1)}{2} = \frac{(y-2)}{4}.$$

These can be written in symmetric form in the following manner

$$\frac{x-1}{2} = \frac{y-2}{4} = \frac{z-4}{-1}.$$

∎

Hence, the normal line is given by the equation above. It is worth noting that the direction numbers $(2, 4, -1)$ apply to the outward-pointing normal at the given point as shown in Figure 4.8. If we redefine $\phi(x, y, z)$ as $\phi(x, y, z) \equiv -x^2 - y^2 + z$, we will then obtain the direction numbers for the inward-pointing normal.

4.5   Divergence and Curl of a Vector Field

The vector character of the operator ∇ suggests that we also consider dot and cross products in which it appears as a factor. Let $\mathbf{F}(x, y, z)$ be a differentiable function, where x, y, z are Cartesian coordinates, and let F_1, F_2, F_3 be the components of $\mathbf{F}$. Then the function

$$\nabla \cdot \mathbf{F} = \left(\mathbf{i}\frac{\partial}{\partial x} + \mathbf{j}\frac{\partial}{\partial y} + \mathbf{k}\frac{\partial}{\partial z} \right) \cdot (F_1\mathbf{i} + F_2\mathbf{j} + F_3\mathbf{k})$$

$$= \frac{\partial F_1}{\partial x} + \frac{\partial F_2}{\partial y} + \frac{\partial F_3}{\partial z} \qquad (4.14)$$

is called the **divergence** of $\mathbf{F}$. This is commonly denoted **div** $\mathbf{F}$. Note that $\nabla \cdot \mathbf{F}$ means the scalar **div** $\mathbf{F}$, whereas ∇f means the vector **grad** f defined by equation (4.7).

In a similar manner, we can define the cross product of ∇ with $\mathbf{F}$ as

$$\nabla \times \mathbf{F} = \begin{vmatrix} \mathbf{i} & \mathbf{j} & \mathbf{k} \\ \frac{\partial}{\partial x} & \frac{\partial}{\partial y} & \frac{\partial}{\partial z} \\ F_1 & F_2 & F_3 \end{vmatrix}$$

$$= \left(\frac{\partial F_3}{\partial y} - \frac{\partial F_2}{\partial z} \right) \mathbf{i} + \left(\frac{\partial F_1}{\partial z} - \frac{\partial F_3}{\partial x} \right) \mathbf{j}$$

$$+ \left(\frac{\partial F_2}{\partial x} - \frac{\partial F_1}{\partial y} \right) \mathbf{k}. \tag{4.15}$$

This expression is known as the **curl** of **F** or the curl of the vector field defined by **F**.

4.6   Physical Interpretation of Divergence

Consider a region of space filled with a moving fluid and let

$$\mathbf{V} = v_1 \mathbf{i} + v_2 \mathbf{j} + v_3 \mathbf{k}$$

be a vector function representing at each point the instantaneous velocity of the particle of the fluid. If we fix our attention on the infinitesimal volume (Figure 4.9) in the region occupied by the fluid, there will be flow through each of its faces, and as a result the amount of fluid within the element may vary. We can compute the loss of fluid from the element in the time Δt to measure this variation.

The volume of the fluid that passes through one face of the element ΔV in time Δt is approximately equal to the component of the fluid velocity normal to the face times the area of the face times Δt, and the corresponding mass flow is the product of this volume and the density of the fluid ρ.

Therefore, computing the loss of fluid through each face in turn (remembering that because the fluid is not assumed to be incompressible, the density as well as the velocity may vary from point to point), we have

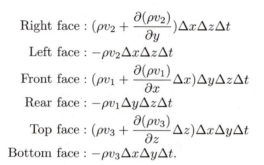

$$\text{Right face} : (\rho v_2 + \frac{\partial(\rho v_2)}{\partial y})\Delta x \Delta z \Delta t$$

$$\text{Left face} : -\rho v_2 \Delta x \Delta z \Delta t$$

$$\text{Front face} : (\rho v_1 + \frac{\partial(\rho v_1)}{\partial x}\Delta x)\Delta y \Delta z \Delta t$$

$$\text{Rear face} : -\rho v_1 \Delta y \Delta z \Delta t$$

$$\text{Top face} : (\rho v_3 + \frac{\partial(\rho v_3)}{\partial z}\Delta z)\Delta x \Delta y \Delta t$$

$$\text{Bottom face} : -\rho v_3 \Delta x \Delta y \Delta t.$$

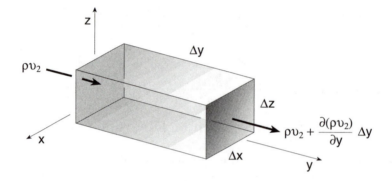

FIGURE 4.9
Typical volume element in a region filled with moving fluid

If these are added and the resulting estimate of absolute loss of fluid from ΔV in the interval Δt is converted into loss per unit volume per unit time dividing by $\Delta V \Delta t \equiv \Delta x \Delta y \Delta z \Delta t$, we obtain the limit:

$$\text{Rate of loss per unit volume} \quad = \frac{\partial(\rho v_1)}{\partial x} + \frac{\partial(\rho v_2)}{\partial y} + \frac{\partial(\rho v_3)}{\partial z} \qquad (4.16)$$

that is precisely the divergence of the vector $(\rho \mathbf{V})$. From fluid mechanics we know that a possible interpretation of the divergence is the rate of loss or gain of fluid per unit volume. The divergence of a vector field can also be interpreted as a measure of the rate of change of the density of the fluid at a point. In other words, **div F** is a measure of the fluid's compressibility. If $\nabla \cdot \mathbf{F} = 0$, the fluid is said to be **incompressible**. In electromagnetic theory, if $\nabla \cdot \mathbf{F} = 0$, the vector field **F** is said to be **solenoidal**.

If the fluid is incompressible, then there is no loss or gain of fluid. In a general element where $\rho = $ constant

$$\nabla \cdot (\rho \mathbf{V}) = 0$$
$$\rho(\nabla \cdot \mathbf{V}) = 0$$
$$\nabla \cdot \mathbf{V} = 0 \qquad (4.17)$$

that is known as the **equation of continuity**.

4.7   Physical Interpretation of the Curl

Consider a body rotating with uniform angular speed $\boldsymbol{\Omega}$ about an axis l (Figure 4.9). We define the **vector angular velocity** $\boldsymbol{\Omega}$ to be a vector of length ω extending along l in the direction in which a right-handed screw would advance if subject to the same rotation as the body. Let $\mathbf{R}$ be the vector drawn from any point O on the axis l to an arbitrary point P in the body. From Figure 4.10, it is evident that the radius at which P rotates is $|\mathbf{R}| \cdot |\sin\theta|$. Therefore, the linear speed of P is

$$|\mathbf{V}| = \Omega|\mathbf{R}| \cdot |\sin\theta| = |\boldsymbol{\Omega}||\mathbf{R}||\sin\theta| = |\boldsymbol{\Omega} \times \mathbf{R}|.$$

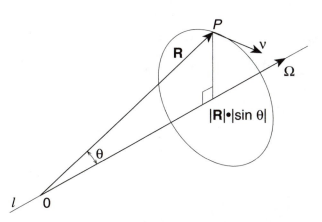

FIGURE 4.10
A Physical interpretation of the curl

Furthermore, the vector velocity $\mathbf{V}$ is directed perpendicular to the plane of $\boldsymbol{\Omega}$ and $\mathbf{R}$ so that $\boldsymbol{\Omega}$, $\mathbf{R}$ and $\mathbf{V}$ form a right-handed system. Hence, the cross product $\boldsymbol{\Omega} \times \mathbf{R}$ gives not only the magnitude of $\mathbf{V}$ but the direction as well. If we take the point O as the origin of coordinates, we can write

$$\mathbf{R} = x\mathbf{i} + y\mathbf{j} + z\mathbf{k} \quad \text{and} \quad \boldsymbol{\Omega} = \Omega_1\mathbf{i} + \Omega_2\mathbf{j} + \Omega_3\mathbf{k}$$

Therefore, the equation $\mathbf{V} = \boldsymbol{\Omega} \times \mathbf{R}$ can be written in the form

$$\mathbf{V} = (\Omega_2 z - \Omega_3 y)\mathbf{i} - (\Omega_1 z - \Omega_3 z)\mathbf{j} + (\Omega_1 y - \Omega_2 x)\mathbf{k}. \qquad (4.18)$$

If we take the curl of $\mathbf{V}$, we have

$$\nabla \times \mathbf{V} = \begin{vmatrix} \mathbf{i} & \mathbf{j} & \mathbf{k} \\ \frac{\partial}{\partial x} & \frac{\partial}{\partial y} & \frac{\partial}{\partial z} \\ (\Omega_2 z - \Omega_3 y) & -(\Omega_1 z - \Omega_3 x) & (\Omega_1 y - \Omega_2 x) \end{vmatrix}.$$

We can expand this, remembering that $\boldsymbol{\Omega}$ is a constant vector, to get

$$\nabla \times \mathbf{V} = 2\Omega_1\mathbf{i} + 2\Omega_2\mathbf{j} + 2\Omega_3\mathbf{k} = 2\boldsymbol{\Omega}$$

$$\boldsymbol{\Omega} = \frac{1}{2}\nabla \times \mathbf{V}. \qquad (4.20)$$

The angular velocity of a uniformly rotating body is equal to one-half the curl of the linear velocity of any point of the body. The aptness of the name *curl* in this connection is apparent.

Example 4.12
Calculate the curl of the vector field if

(a) $\mathbf{F} = 2xy\mathbf{i} + z\mathbf{j} + x^2 \cos(yz)\mathbf{k}$.

(b) $\mathbf{F} = \frac{qQ}{4\pi\epsilon_0|\mathbf{R}|^3}\mathbf{R}$, where $\mathbf{R} = x\mathbf{i} + y\mathbf{j} + z\mathbf{k}$.

⬜

SOLUTION

(a)

$$\nabla \times \mathbf{F} = \begin{vmatrix} \mathbf{i} & \mathbf{j} & \mathbf{k} \\ \frac{\partial}{\partial x} & \frac{\partial}{\partial y} & \frac{\partial}{\partial z} \\ 2xy & z & x^2\cos(yz) \end{vmatrix}$$

$$= (-x^2 z \sin(yz) - 1)\mathbf{i} + (-2x\cos(yz))\mathbf{j} + (-2x)\mathbf{k}.$$

(b)

$$\nabla \times \mathbf{F} = \frac{qQ}{4\pi\epsilon_0} \begin{vmatrix} \mathbf{i} & \mathbf{j} & \mathbf{k} \\ \frac{\partial}{\partial x} & \frac{\partial}{\partial y} & \frac{\partial}{\partial z} \\ \frac{x}{|\mathbf{R}|^3} & \frac{y}{|\mathbf{R}|^3} & \frac{z}{|\mathbf{R}|^3} \end{vmatrix}.$$

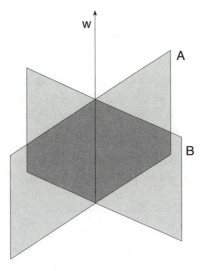

FIGURE 4.11
Paddle device

The x-component of $\nabla \times \mathbf{F}$ is $qQ/(4\pi\epsilon_0)$ multiplied by

$$\frac{\partial}{\partial y}\left(\frac{z}{|\mathbf{R}|^3}\right) - \frac{\partial}{\partial z}\left(\frac{y}{|\mathbf{R}|^3}\right) = \frac{\partial}{\partial y}\left(\frac{z}{(x^2 + y^2 + z^2)^{3/2}}\right)$$

$$- \frac{\partial}{\partial z}\left(\frac{y}{(x^2 + y^2 + z^2)^{3/2}}\right)$$

$$= \frac{-3yz}{(x^2 + y^2 + z^2)^{5/2}}$$

$$+ \frac{3yz}{(x^2 + y^2 + z^2)^{5/2}} = 0.$$

Similar results for the y- and z-components give $\nabla \times \mathbf{F} = \mathbf{0}$.

∎

If a vector $\rho\mathbf{V}$ is defined as above for gas flow through a region D, then the curl of $\rho\mathbf{V}$ describes the tendency for the motion of the gas to be circular rather than flowing in a straight line. This suggests why the term **curl** is used. With this interpretation, the following definition seems reasonable.

DEFINITION 4.3 *A vector field* $\mathbf{F}$ *is said to be irrotational in a region* D *if in* D

$$\nabla \times \mathbf{F} = \mathbf{0}. \tag{4.21}$$

The word *curl* was introduced by Maxwell in his studies of electromagnetic fields. However, the curl is easily understood in connection with the flow of fluids. If a paddle device (Figure 4.11) is inserted in a flowing fluid, then the curl of the velocity field $\mathbf{F}$ is a measure of the tendency of the fluid to turn the device about its vertical axis w. If $\nabla \times \mathbf{F} = \mathbf{0}$, then the flow of the fluid is **irrotational** (from Definition 4.3), which means that it is free of vortices or whirlpools that would cause the paddle to rotate.

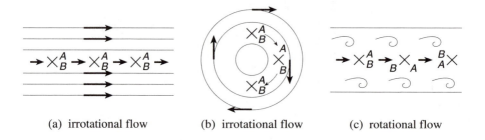

(a) irrotational flow   (b) irrotational flow   (c) rotational flow

FIGURE 4.12
Fluid flow

For this reason, in some science texts, the word *rotation* is sometimes used instead of *curl*. The symbol curl $\mathbf{F}$ is then replaced by rot $\mathbf{F}$. In Figure 4.12 the axis w points straight out of the page. Note from Figure 4.12 (b) that **irrotational** does *not* mean that the fluid does not rotate.

4.8   The Laplacian Operator

In the applied field of electrostatics, the gradient of the potential is a scalar multiple of the electric field intensity. The divergence of the electric intensity is related to the charge density. In light of this and other considerations, it is convenient to introduce a single operator that is the composite of the gradient (**grad**) and the divergence (**div**). This operator of a scalar field ϕ is defined to be **div** (**grad** ϕ). It must be noted that **grad** ϕ is a vector field, while the divergence of **grad** ϕ is a scalar field. The Laplacian of ϕ is often referred to as "del squared ϕ". This can be written in the following manner, using $\nabla^2 \phi$, for simplicity

$$\nabla^2 \phi = \nabla \cdot (\nabla \phi) = \frac{\partial^2 \phi}{\partial x^2} + \frac{\partial^2 \phi}{\partial y^2} + \frac{\partial^2 \phi}{\partial z^2}$$

because we have

$$\nabla \cdot (\nabla \phi) = \nabla \cdot \left(\frac{\partial \phi}{\partial x} \mathbf{i} + \frac{\partial \phi}{\partial y} \mathbf{j} + \frac{\partial \phi}{\partial z} \mathbf{k} \right) = \frac{\partial^2 \phi}{\partial x^2} + \frac{\partial^2 \phi}{\partial y^2} + \frac{\partial^2 \phi}{\partial z^2}.$$

From the above, we see that the symbol ∇^2 can be considered an abbreviation for

$$\frac{\partial^2}{\partial x^2} + \frac{\partial^2}{\partial y^2} + \frac{\partial^2}{\partial z^2}.$$

This equation occurs so often that we abbreviate it as $\nabla^2 \phi$. The operator

$$\nabla^2 = \nabla \cdot \nabla = |\nabla^2|$$

is called the **Laplace operator** because of its relation to **Laplace's equation**

$$\nabla^2 \phi = \frac{\partial^2 \phi}{\partial x^2} + \frac{\partial^2 \phi}{\partial y^2} + \frac{\partial^2 \phi}{\partial z^2} = 0.$$

Thus, the Laplacian of ϕ is just the sum of the second partial derivatives of ϕ. The Laplacian can also act on a vector quantity. In this case, the result is a vector whose components are the Laplacians of the components of the original vector

$$\nabla^2 \mathbf{U} = \left(\nabla^2 u_1, \nabla^2 u_2, \nabla^2 u_3 \right).$$

It can further be stated that any function satisfying this equation in a given region is said to be *harmonic* in that region. For example, in any region where the charge density is zero, the electric potential of a static distribution of charges is harmonic. A function describing the steady-state temperature distribution of a homogeneous material is harmonic in the interior of the region occupied by the material. Laplace's equation is seen to occur mainly in gravitation, electrostatics, steady-state heat flow, and fluid flow. The Laplacian operator is by far the most important differential operator in mathematical physics.

If ϕ is a scalar, then $\nabla^2 \phi(x, y, z)$ denotes the value of $\nabla^2 \phi$ at the point (x, y, z). This is a number that tells us something about the behavior of the scalar field in the neighborhood of (x, y, z). Roughly speaking, it provides a measure of the *difference between the average value of the field in the immediate neighborhood of the point and the precise value of the field at the point*. It must be noted that the word "average" here refers to an average

---

over a region of space, not to a time average. Variations with time t do not enter into the calculation of $\nabla^2\phi$.

Thus, if $\nabla^2\phi$ is positive at a point, and ϕ denotes the temperature, this means that the temperature in the vicinity of the point is, on average, greater than the temperature at the point itself. In particular, if the temperature takes its minimum value at a certain point in space, it is reasonable to expect that the value of $\nabla^2\phi$ will *not* be negative at that point. From this viewpoint, the Laplacian can be viewed as a sort of three-dimensional generalization of the ordinary differential operator d^2/dx^2, which is used in elementary calculus to test extreme points to see if they represent maxima or minima.

4.9   Vector Identities

Although we continue to use the del notation, formally manipulating

$$\nabla = \mathbf{i}\frac{\partial}{\partial x} + \mathbf{j}\frac{\partial}{\partial y} + \mathbf{k}\frac{\partial}{\partial z}$$

as though it were a vector, this practice has certain hazards. Keep in mind that the derivative operators appearing in the del operator act only on functions appearing to the right of the del operator. For example, suppose that

$$\mathbf{F} = x^3 y\mathbf{i} + y^2\mathbf{j} + x^2 z\mathbf{k}, \quad \mathbf{R} = x\mathbf{i} + y\mathbf{j} + z\mathbf{k}.$$

Let us now compare the expressions $(\nabla\cdot\mathbf{R})\mathbf{F}$ and $(\mathbf{R}\cdot\nabla)\mathbf{F}$. For the first expression we have

$$(\nabla\cdot\mathbf{R})\mathbf{F} = 3\mathbf{F} = 3x^3 y\mathbf{i} + 3y^2\mathbf{j} + 3x^2 z\mathbf{k}.$$

On the other hand, in the second expression, $\mathbf{R}$ is to the left of ∇, and therefore the derivatives in the del operator do not act on $\mathbf{R}$. In this case we have

$$(\mathbf{R}\cdot\nabla)\mathbf{F} = \left(x\frac{\partial}{\partial x} + y\frac{\partial}{\partial y} + z\frac{\partial}{\partial z}\right)(x^3 y\mathbf{i} + y^2\mathbf{j} + x^2 z\mathbf{k})$$
$$= x(3x^2 y\mathbf{i} + 2xz\mathbf{k}) + y(x^3\mathbf{i} + 2y\mathbf{j}) + z(x^2\mathbf{k})$$
$$= 4x^3 y\mathbf{i} + 2y^2\mathbf{j}3x^2 z\mathbf{k}.$$

To further confuse matters, parentheses are sometimes omitted so that the reader should supply such parentheses as necessary to make the expression meaningful. For example, $\nabla\cdot\mathbf{R}\mathbf{F}$ and $\mathbf{R}\cdot\nabla\mathbf{F}$ must mean $(\nabla\cdot\mathbf{R})\mathbf{F}$ and $(\mathbf{R}\cdot$

$\nabla)\mathbf{F}$, respectively, because $\nabla \cdot (\mathbf{R}\mathbf{F})$ and $\mathbf{R} \cdot (\nabla\mathbf{F})$ do not make sense within the framework of vector analysis. Similarly, $\nabla \cdot f\mathbf{F}$ means $\nabla \cdot (f\mathbf{F})$, simply the divergence of $f\mathbf{F}$, because $\nabla \cdot f$, and hence $(\nabla \cdot f)\mathbf{F}$, is meaningless. Because of these problems of interpretation, it is vitally important to preserve the working order with ∇. We now list a number of identities. Here $\mathbf{U}$ and $\mathbf{V}$ denote vector fields, while ϕ denotes a scalar field.

$$\nabla \cdot \phi\mathbf{V} = \phi\nabla \cdot \mathbf{V} + \mathbf{V} \cdot \nabla\phi \tag{4.22}$$
$$\nabla \times \phi\mathbf{V} = \phi\nabla \times \mathbf{V} + (\nabla\phi) \times \mathbf{V} \tag{4.23}$$
$$\nabla \cdot (\mathbf{U} \times \mathbf{V}) = \mathbf{V} \cdot \nabla \times \mathbf{U} - \mathbf{U} \cdot \nabla \times \mathbf{V} \tag{4.24}$$
$$\nabla \times (\mathbf{U} \times \mathbf{V}) = (\mathbf{V} \cdot \nabla)\mathbf{U} - (\mathbf{U} \cdot \nabla)\mathbf{V}$$
$$+ \mathbf{U}(\nabla \cdot \mathbf{V})$$
$$- \mathbf{V}(\nabla \cdot \mathbf{U}) \tag{4.25}$$
$$\nabla(\mathbf{U} \cdot \mathbf{V}) = (\mathbf{U} \cdot \nabla)\mathbf{V} + (\mathbf{V} \cdot \nabla)\mathbf{U}$$
$$+ \mathbf{U} \times (\nabla \times \mathbf{V}) + \mathbf{V} \times (\nabla \times \mathbf{U}) \tag{4.26}$$
$$\nabla \times \nabla\phi = \mathbf{0} \tag{4.27}$$
$$\nabla \cdot \nabla \times \mathbf{V} = 0 \tag{4.28}$$
$$\nabla \times (\nabla \times \mathbf{V}) = \nabla(\nabla \cdot \mathbf{V}) - \nabla \cdot \nabla\mathbf{V} = \nabla(\nabla \cdot \mathbf{V}) - \nabla^2\mathbf{V} \tag{4.29}$$

These identities can all be verified by direct expansion. For example, to prove (4.22), we have

$$\nabla \cdot (\phi\mathbf{V}) = \nabla \cdot (\phi(v_1\mathbf{i} + v_2\mathbf{j} + v_3\mathbf{k}))$$
$$= \frac{\partial(\phi v_1)}{\partial x} + \frac{\partial(\phi v_2)}{\partial y} + \frac{\partial(\phi v_3)}{\partial z}$$
$$= (\phi\frac{\partial v_1}{\partial x} + v_1\frac{\partial \phi}{\partial x}) + (\phi\frac{\partial v_2}{\partial y} + v_2\frac{\partial \phi}{\partial y}) + (\phi\frac{\partial v_3}{\partial z} + v_3\frac{\partial \phi}{\partial z})$$
$$= \phi(\frac{\partial v_1}{\partial x} + \frac{\partial v_2}{\partial y} + \frac{\partial v_3}{\partial z}) + v_1\frac{\partial \phi}{\partial x} + v_2\frac{\partial \phi}{\partial y} + v_3\frac{\partial \phi}{\partial z}$$
$$= \phi\nabla \cdot \mathbf{V} + \mathbf{V} \cdot \nabla\phi$$

To establish these formulae, it is generally easier if ∇ is treated as a vector. This leads to the manipulation of the expressions, according to the appropriate formulae from vector algebra, and finally giving ∇ its operational meaning. Because ∇ is a linear combination of scalar differential operators that obey the usual product rule of differentiation, it is clear that ∇ itself has this property.

We can apply ∇ to products of various sorts by assuming that each of the factors, in turn, is the only one that is variable and adding the partial results so obtained. As a notation in helping us to determine these partial results, it is helpful to attach to ∇, whenever it is followed by more than one factor, a subscript indicating the one factor upon which it is currently allowed to operate.

To prove (4.23), we first suppose that the scalar function ϕ is constant. Then we can write

$$\nabla_v \times (\phi \mathbf{V}) = \phi \nabla \times \mathbf{V}.$$

Similarly, if we regard $\mathbf{V}$ as constant and ϕ as variable, we have

$$\nabla_\phi \times (\phi \mathbf{V}) = (\nabla \phi) \times \mathbf{V}$$

where the parentheses now restrict the effect of ∇ to the factor ϕ alone. Adding these two partial results gives

$$\nabla_v \times (\phi \mathbf{V}) + \nabla_\phi \times (\phi \mathbf{V}) \equiv \nabla \times (\phi \mathbf{V}) = \phi \nabla \times \mathbf{V} \\ + (\nabla \phi) \times \mathbf{V}.$$

To prove (4.24), from the cyclic properties of scalar triple products, we have the following

$$\nabla_u \cdot (\mathbf{U} \times \mathbf{V}) = \nabla \cdot \mathbf{V} \times \mathbf{U} \quad \text{and}$$
$$\nabla_v \cdot (\mathbf{U} \times \mathbf{V}) = -\mathbf{U} \cdot \nabla \times \mathbf{V}.$$

Therefore, adding these two partial results gives

$$\nabla_u \cdot (\mathbf{U} \times \mathbf{V}) + \nabla_v (\mathbf{U} \times \mathbf{V}) \equiv \nabla \cdot (\mathbf{U} \times \mathbf{V}) \\ = \mathbf{V} \cdot \nabla \times \mathbf{U} - \mathbf{U} \cdot \nabla \times \mathbf{V}.$$

To prove (4.25), we have

$$\nabla_u \times (\mathbf{U} \times \mathbf{V}) = (\nabla_u \cdot \mathbf{V})\mathbf{U} - (\nabla_u \cdot \mathbf{U})\mathbf{V} = \mathbf{V} \cdot \nabla \mathbf{U} - \mathbf{V} \nabla \cdot \mathbf{U}$$

and

$$\nabla_v \times (\mathbf{U} \times \mathbf{V}) = (\nabla_v \cdot \mathbf{V})\mathbf{U} - (\nabla_v \cdot \mathbf{U})\mathbf{V} = \mathbf{U} \nabla \cdot \mathbf{V} - \mathbf{U} \cdot \nabla \mathbf{V}.$$

Adding gives

$$\nabla_u \times (\mathbf{U} \times \mathbf{V}) + \nabla_v \times (\mathbf{U} \times \mathbf{V}) \equiv \nabla \times (\mathbf{U} \times \mathbf{V})$$
$$= \mathbf{V} \cdot \nabla \mathbf{U} - \mathbf{U} \cdot \nabla \mathbf{V}$$
$$+ \mathbf{U}\nabla \cdot \mathbf{V} - \mathbf{V}\nabla \cdot \mathbf{U}.$$

To prove (4.26), we note that

$$\mathbf{U} \times (\nabla \times \mathbf{V}) \equiv \mathbf{U} \times (\nabla_v \times \mathbf{V}) = (\mathbf{U} \cdot \mathbf{V})\nabla_v - (\mathbf{U} \cdot \nabla)\mathbf{V}$$
$$= \nabla_v(\mathbf{U} \cdot \mathbf{V}) - \mathbf{U} \cdot \nabla \mathbf{V}$$

and

$$\mathbf{V} \times (\nabla \times \mathbf{U}) \equiv \mathbf{V} \times (\nabla_u \times \mathbf{U}) = (\mathbf{V} \cdot \mathbf{U})\nabla_u - (\mathbf{V} \cdot \nabla)\mathbf{U}$$
$$= \nabla_u(\mathbf{U} \cdot \mathbf{V}) - \mathbf{V} \cdot \nabla \mathbf{U}.$$

Hence, transposing and adding, we find

$$\nabla_u(\mathbf{U} \cdot \mathbf{V}) + \nabla_v(\mathbf{U} \cdot \mathbf{V}) \equiv \nabla(\mathbf{U} \cdot \mathbf{V})$$
$$= \mathbf{U} \times (\nabla \times \mathbf{V}) + \mathbf{V} \times (\nabla \times \mathbf{U})$$
$$+ \mathbf{U} \cdot \nabla \mathbf{V} + \mathbf{V} \cdot \nabla \mathbf{U}.$$

The fact that the operational coefficient $\nabla \times \nabla$ appears in (4.27) as the cross product of identical factors suggests that $\nabla \times \nabla\phi$ is zero and can be easily verified. Similarly, the fact that the left number of formula (4.28) appears as a scalar triple product with two identical factors suggests the truth of the formula, and again it is not difficult to show that $\nabla \cdot \nabla \times \mathbf{V}$ is always equal to zero.

Formula (4.29) can be established either by direct expansion or it can be verified by applying the usual rule for expanding a vector triple product.

$$\nabla \times (\nabla \times \mathbf{V}) = (\nabla \cdot \mathbf{V})\nabla - (\nabla \cdot \nabla)\mathbf{V} = \nabla(\nabla \cdot \mathbf{V}) - \nabla^2\mathbf{V}$$

where the conventional symbol ∇^2 has been substituted for the second-order operator

$$\nabla \cdot \nabla = \left(\mathbf{i}\frac{\partial}{\partial x} + \mathbf{j}\frac{\partial}{\partial y} + \mathbf{k}\frac{\partial}{\partial z}\right) \cdot \left(\mathbf{i}\frac{\partial}{\partial x} + \mathbf{j}\frac{\partial}{\partial y} + \mathbf{k}\frac{\partial}{\partial z}\right)$$
$$= \frac{\partial^2}{\partial x^2} + \frac{\partial^2}{\partial y^2} + \frac{\partial^2}{\partial z^2}.$$

Example 4.13
Calculate $\nabla \cdot \mathbf{R}$, $\nabla \cdot \mathbf{R} f(r)$ and then show that

$$\nabla \cdot \mathbf{R} r^{n-1} = (n+2) r^{n-1}$$

Note that here the magnitude of the position vector is denoted as r, i.e., $|\mathbf{R}| = r$.   ▯

SOLUTION   We know that

$$\nabla \cdot \mathbf{R} = \left(\mathbf{i} \frac{\partial}{\partial x} + \mathbf{j} \frac{\partial}{\partial y} + \mathbf{k} \frac{\partial}{\partial z} \right) \cdot (\mathbf{i}x + \mathbf{j}y + \mathbf{k}z)$$

$$= \frac{\partial x}{\partial x} + \frac{\partial y}{\partial y} + \frac{\partial z}{\partial z},$$

or $\nabla \cdot \mathbf{R} = 3.$

For the second part, we have

$$\nabla \cdot \mathbf{R}\, f(r) = \frac{\partial}{\partial x}[x f(r)] + \frac{\partial}{\partial y}[y f(r)] + \frac{\partial}{\partial x}[z f(r)]$$

$$= 3f(r) + \frac{x^2}{r}\frac{df}{dr} + \frac{y^2}{r}\frac{df}{dr} + \frac{z^2}{r}\frac{df}{dr}$$

$$= 3f(r) + r\frac{df}{dr}.$$

Now to show the identity, we make use of the above taking $f(r) = r^{n-1}$,

$$\nabla \cdot \mathbf{R}\, r^{n-1} = \nabla \cdot \mathbf{R}_0\, r^n$$

$$= 3r^{n-1} + (n-1)r^{n-1}$$

$$= (n+2)r^{n-1}.$$

▮

It is worth noting that this divergence vanishes for $n = -2$.

Exercises

1. Find the directional derivative of f at the given point in the direction indicated by the angle θ.

   (a) $f(x, y) = x^2 y^3 + 2x^4 y$, $(1, -2)$, $\theta = \pi/3$
   (b) $f(x,) = (x^2 - y)^3$ $(3, 1)$, $\theta = 3\pi/4$

(d) $f(x,) = y^x$, $(1,2)$, $\theta = \pi/2$

(d) $f(x,) = \sin(x+2y)$, $(4,-2)$, $\theta = -2\pi/3$.

For the following problems,

(a) find the gradient of f

(b) evaluate the gradient at the point P

(c) find the rate of change of f at P in the direction of the vector **U**.

2. $f(x,y) = x^3 - 4x^2y + y^2$, $P(0,-1)$, $\mathbf{U} = (3/5, 4/5)$.

3. $f(x,y) = e^x \sin y$, $P(1, \pi/4)$, $\mathbf{U} = (-1/\sqrt{5}, 2/\sqrt{5})$.

4. $f(x,y,z) = xy^2z^3$, $P(1,-2,1)$, $\mathbf{U} = (1/\sqrt{3}, -1/\sqrt{3}, 1/\sqrt{3})$.

5. $f(x,y,z) = xy + yz^2 + xz^3$, $P(2,0,3)$, $\mathbf{U} = (-2/3, -1/3, 2/3)$.

6. Find the directional derivative of the function at the given point in the direction of the vector **V**

  (a) $f(x,y) = \sqrt{x-y}$, $(5,1)$, $\mathbf{V} = (12,5)$
  (b) $f(x,y) = xe^{xy}$ $(-3,0)$, $\mathbf{V} = 2\mathbf{i} + 3\mathbf{j}$
  (c) $f(x,y,z) = \sqrt{xyz}$ $(2,4,2)$, $\mathbf{V} = (4,2,-4)$
  (d) $f(x,y) = x\tan^{-1}(y/z)$ $(1,2-2)$, $\mathbf{V} = \mathbf{i} + \mathbf{j} - \mathbf{k}$
  (e) $f(x,y) = xe^{yz} + xye^z$ $(-2,1,1)$, $\mathbf{V} = \mathbf{i} - 2\mathbf{j} + 3\mathbf{k}$.

7. Find the maximum rate of change of f at the given point and the direction in which it occurs.

  (a) $f(x,y) = xe^{-y} + 3y$, $(1,0)$
  (b) $f(x,y) = \ln(x^2 + y^2)$, $(1,2)$
  (c) $f(x,y) = \sqrt{x^2 + 2y}$, $(4,10)$
  (d) $f(x,y) = x + y/z$, $(4,3,-1)$
  (e) $f(x,y) = \cos(3x + 2y)$, $(\pi/6, -\pi/8)$
  (f) $f(x,y) = x/y + y/z$, $(4,2,1)$.

8. The temperature at a point (x,y,z) is given by

$$T(x,y,z) = 200e^{-x^2-3y^2-9z^2}$$

where T is measured in degrees Celsius and x, y, z in meters.

  (a) Find the rate of change of temperature at the point $P(2,-1,2)$ in the direction toward the point $(3,-3,3)$.

(b) In which direction does the temperature increase fastest at P?

(c) Find the maximum rate of increase of T.

9. Suppose that you are climbing a hill whose shape is given by equation $z = 1000 - 0.01x^2 - 0.02y^2$, and you are standing at a point with coordinates $(60, 100, 764)$.

    (a) In which direction should you initially proceed in order to reach the top of the hill fastest?

    (b) If you climb in that direction, at what angle above the horizontal will you be initially climbing?

10. Given $\mathbf{F}(x, y, z) = x^2 y \mathbf{i} + z \mathbf{j} - (x + y - z)\mathbf{k}$ find

    (a) $\nabla \cdot \mathbf{F}$

    (b) $\nabla \times \mathbf{F}$

    (c) $\nabla(\nabla \cdot \mathbf{F})$.

11. Which of the following functions satisfy Laplace's equation?

    (a) $f(x, y, z) = e^z \sin y$

    (b) $f(x, y, z) = \sin x \sinh y \cos \cosh z$

    (c) $f(x, y, z) = \sin px \sinh qy$ (p and q are constants).

12. Given that $f(x, y, z) = 2x^2 + y$ and $\mathbf{R} = x\mathbf{i} + y\mathbf{j} + z\mathbf{k}$, find

    (a) ∇f

    (b) $\nabla \cdot \mathbf{R}$

    (c) $\nabla^2 f$

    (d) $\nabla \times (f\mathbf{R})$.

13. Evaluate $\nabla^2[(\mathbf{i} + \mathbf{j} + \mathbf{k}) \times \nabla(\mathbf{R} \cdot \mathbf{R})^2]$ where $\mathbf{R} = x\mathbf{i} + y\mathbf{j} + z\mathbf{k}$.

14. Evaluate $\mathbf{A} \cdot \nabla \mathbf{R} + \nabla(\mathbf{A} \cdot \mathbf{R}) + \mathbf{A} \cdot \nabla \times \mathbf{R}$ where $\mathbf{A}$ is a constant vector field and $\mathbf{R} = x\mathbf{i} + y\mathbf{j} + z\mathbf{k}$.

15. If $r^2 = x^2 + y^2 + z^2$, $\mathbf{R} = x\mathbf{i} + y\mathbf{j} + z\mathbf{k}$ and $\mathbf{A}$ is a constant vector field, find

    (a) $\nabla \cdot (r^2 \mathbf{A})$

    (b) $\nabla \times (r^2 \mathbf{A})$

    (c) $\mathbf{R} \cdot \nabla(r^2 \mathbf{A})$

    (d) $\nabla(\mathbf{A} \cdot \mathbf{R})^4$

    (e) $\nabla \cdot (r\mathbf{A})$

(f) $\mathbf{R} \cdot \nabla (\mathbf{A} \cdot \mathbf{RA})$

(g) $\nabla \cdot (\mathbf{A} \times \mathbf{R})$

(h) $\nabla^2 (\mathbf{R} \cdot \mathbf{R})$.

16. If r is the distance from the origin to the point (x, y, z) and $\mathbf{A}$ is a constant vector, evaluate

$$\nabla \left(\mathbf{A} \cdot \nabla \frac{1}{r} \right) + \nabla \times \left(\mathbf{A} \times \nabla \frac{1}{r} \right).$$

17. Given $\phi = \tan^{-1} x + \tan^{-1} y$ and $\psi = \frac{x+y}{1-xy}$, show that

$$\nabla \phi \times \nabla \psi = 0.$$

18. Find the divergence of the field

$$\mathbf{F} = \frac{x\mathbf{i} + y\mathbf{j} + z\mathbf{k}}{(x^2 + y^2 + z^2)^{3/2}}.$$

Is the divergence of this field defined at every point in space?

19. Given that $\mathbf{R} = x\mathbf{i} + y\mathbf{j} + z\mathbf{k}$ and $r = |\mathbf{R}| = \sqrt{x^2 + y^2 + z^2}$, prove that

$$\nabla r^n = nr^{n-2}\mathbf{R}.$$

20. Using the same definition of r as in the previous question, show that

$$\nabla \times \left[\frac{1}{r} (\mathbf{A} \times \mathbf{R}) \right] = \frac{1}{r}\mathbf{A} + \frac{\mathbf{A} \cdot \mathbf{R}}{r^3}\mathbf{R}$$

for any constant vector $\mathbf{A}$.

21. (a) Show that a necessary and sufficient condition that $u(x, y, z)$ and $v(x, y, z)$ are related by some function $f(u, v) = 0$ is that $(\nabla u) \times (\nabla v) = 0$.

(b) If $u = u(x, y)$ and $v = v(x, y)$, show that the condition $(\nabla u) \times (\nabla v) = 0$ leads to the two-dimensional Jacobian

$$J \left(\frac{u, v}{x, y} \right) = \begin{vmatrix} \frac{\partial u}{\partial x} & \frac{\partial u}{\partial y} \\ \frac{\partial v}{\partial x} & \frac{\partial v}{\partial y} \end{vmatrix} = 0.$$

The functions u and v are assumed differentiable.

22. If a force **F** is given by

$$\mathbf{F} = (x^2 + y^2 + z^2)^n (\mathbf{i}x + \mathbf{j}y + \mathbf{k}z),$$

find

(a) $\nabla \cdot \mathbf{F}$,

(b) $\nabla \times \mathbf{F}$,

(c) A scalar potential $\phi(x, y, z)$ so that $\mathbf{F} = -\nabla\phi$.

(d) For what value of the exponent n does the scalar potential diverge at both the origin and infinity?

[Hint: When a vector **B** is solenoidal, i.e., $\nabla \cdot \mathbf{B} = 0$, a vector potential exists such that $\mathbf{B} = \nabla \times \mathbf{A}$.]

23. The origin of the Cartesian coordinates is at the Earth's center. The moon is on the z-axis, a fixed distance R away (center-to-center distance). The tidal force exerted by the moon on a particle at the earth's surface (point x, y, z) is given by

$$F_x = -GMm\frac{x}{R^3}, \quad F_y = -GMm\frac{y}{R^3} \quad F_z = -GMm\frac{z}{R^3}.$$

Find the potential that yields this tidal force.

24. Vector **B** is formed by the product of two gradients

$$\mathbf{B} = (\nabla u) \times (\nabla v)$$

where u and v are scalar functions.

(a) Show that **B** is solenoidal.

(b) Show that

$$\mathbf{A} = \frac{1}{2}(u\nabla v - v\nabla u)$$

is a vector potential for **B** where

$$\mathbf{B} = \nabla \times \mathbf{A}.$$

Chapter 5

Line, Surface and Volume Integrals

5.1 Introduction

The notion of the definite integral $\int_a^b f(x)dx$, i.e., integration of a function defined over an interval, can be generalized to integration of a function defined along a curve. With this in mind, we now introduce some terminology about curves. Suppose C is a curve parameterized by $x = f(t)$, $y = g(t)$, $a \le t \le b$ and A and B are the points $(f(a), g(a))$ and $(f(b), g(b))$, respectively. We say that:

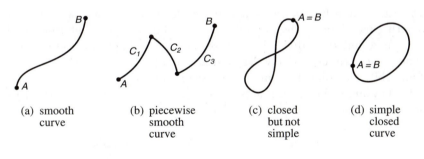

    (a) smooth   (b) piecewise   (c) closed   (d) simple
      curve   smooth   but not   closed
                  curve   simple   curve

FIGURE 5.1
Categories of curves

1. C is a **smooth curve** if f' and g' are continuous on the closed interval $[a, b]$ and not simultaneously zero on the open interval (a, b).

2. C is **piecewise smooth** if it consists of a finite number of smooth curves $C_1, C_2, \ldots, C_n$ joined end to end — i.e., $C = C_1 \cup C_2 \cup \ldots \cup C_n$.

3. C is a **closed curve** if $A = B$.

4. C is a **simple closed curve** if $A = B$ and the curve does not cross itself.

5. If C is not a closed curve, then the **positive direction** on C is the direction corresponding to increasing values of t.

The graphs in Figure 5.1 (a), (b), (c) and (d) illustrate each type of curve defined in $1 - 5$.

This same terminology carries over in a natural manner to curves in space. For example, a curve C defined by $x = f(t)$, $y = g(t)$, $z = h(t)$, $a \le t \le b$, is smooth if f', g' and h' are continuous on $[a, b]$ and not simultaneously zero.

If we partition the interval $[a, b]$ into n sub-intervals $[x_{k-1}, x_k]$ of length $\Delta x_k = x_k - x_{k-1}$, we can denote this partition to be P. We can then take $\|P\|$ to be the length of the longest sub-interval and call it the **norm** of the partition P. Then from calculus we can define the definite integral of a single variable to be given by the limit of the sum:

$$\int_a^b f(x)dx = \lim_{\|P\| \to 0} \sum_{k=1}^n f(x_k)\Delta x_k. \tag{5.1}$$

Alternatively, we could let Δs_i be the length of each sub-arc of the curve. Then the limit shown in equation (5.1) would become

$$\int_a^b f(x)dx = \lim \Delta s_i \to 0 \sum_{k=1}^n f(x_k)\Delta x_k. \tag{5.2}$$

Let $F(x, y)$ be a function of x and y, and let C be a continuous, piecewise smooth curve joining the points A and B. Furthermore, let the arc of C between A and B be divided into n segments Δs_i whose projections on the x and y axes are respectively Δx_i and Δy_i and let (ξ_i, η_i) be the coordinates of an arbitrary point in segment Δs_i (see Figure 5.2).

If we evaluate the given function $F(x, y)$ at each of the points (ξ_i, η_i) and form the products

$$F(\xi_i, \eta_i)\Delta x_i \quad F(\xi_i, \eta_i)\Delta y_i \quad F(\xi_i, \eta_i)\Delta s_i$$

and then sum over all the subdivisions of the arc AB, we have the three sums

$$\sum_{i=1}^n F(\xi_i, \eta_i)\Delta x_i \quad \sum_{i=1}^n F(\xi_i, \eta_i)\Delta y_i \quad \sum_{i=1}^n F(\xi_i, \eta_i)\Delta s_i.$$

The limits of these sums, as n becomes infinite in such a way that the length of each Δs_i approaches zero, are known as **line integrals** and are written, respectively,

$$\int_C F(x, y)dx \qquad \int_C F(x, y)dy \qquad \int_C F(x, y)ds.$$

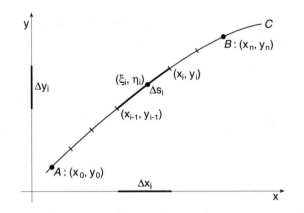

FIGURE 5.2
Subdivision of an arc preparatory to defining a line integral

A more appropriate name might be curvilinear integral instead of line integral. The word "line" is regarded as meaning "curved line" as opposed to a straight line. The following definition summarizes the preceding points.

DEFINITION 5.1 *Let G be a function of two variables x and y defined on a region of the plane containing a smooth curve C.*

1. *The line integral of G along C from A to B with respect to x is*

$$\int_C G(x,y)dx = \lim_{\|P\|\to 0} \sum_{k=1}^{n} G(x_k^*, y_k^*)\Delta x_k.$$

2. *The line integral of G along C from A to B with respect to y is*

$$\int_C G(x,y)dy = \lim_{\|P\|\to 0} \sum_{k=1}^{n} G(x_k^*, y_k^*)\Delta y_k.$$

3. *The line integral of G along C from A to B with respect to arc length is*

$$\int_C G(x,y)ds = \lim_{\|P\|\to 0} \sum_{k=1}^{n} G(x_k^*, y_k^*)\Delta S_k.$$

It can be shown that if $G(x,y)$ is continuous on C, then the integrals defined in 1, 2 and 3 exist. We shall assume continuity as a matter of course.

In these definitions, Δx_i and Δy_i are signed quantities whereas ΔS_i is intrinsically positive. Thus, the following properties of ordinary definite integrals hold

(a) $\int_A^B c\phi(t)dt = c\int_A^B \phi(t)dt$, c a constant

(b) $\int_A^B [\phi_1(t) \pm \phi_2(t)]dt = \int_A^B \phi_1(t)dt \pm \int_A^B \phi_2(t)dt$

(c) $\int_A^B \phi(t)dt = -\int_B^A \phi(t)dt$

(d) $\int_A^P \phi(t)dt + \int_P^B \phi(t)dt = \int_A^B \phi(t)dt$.

are equally valid for line integrals of the first two types 1 and 2, provided that throughout each formula the curve joining A and B remains the same. On the other hand, line integrals of the third type 3, although they do have properties (a) and (b), do not have property (c) because, in fact

$$\int_A^B F(x,y)ds = \int_B^A F(x,y)ds.$$

Moreover, property (d) holds for these integrals if and only if P is between A and B on the path of integration.

5.1.0.1 Method of Evaluation — Curve Defined Parametrically

The line integrals in Definition 5.1 can be evaluated in two ways, depending on whether the curve C is defined parametrically or by an explicit function. In either case, the basic idea is to convert the line integral to a definite integral in a single variable. If C is a smooth curve parameterized by $x = f(t)$, $y = g(t)$, $a \leq t \leq b$, then we simply replace x and y in the integral by the functions $f(t)$ and $g(t)$, and the appropriate differential dx, dy or ds by $f'dt$, $g'dt$, or $\sqrt{(f'(t))^2 + (g'(t))^2}dt$. The expression

$$ds = \sqrt{(f'(t))^2 + (g'(t))^2}dt$$

is called the **differential of arc length**. The integration is carried out with respect to the variable t in the usual manner:

$$\int_C G(x,y)dx = \int_a^b G(f(t),g(t))f'(t)dt \tag{5.3}$$

$$\int_C G(x,y)dy = \int_a^b G(f(t),g(t))g'(t)dt \tag{5.4}$$

$$\int_C G(x,y)ds = \int_a^b G(f(t),g(t))\sqrt{(f'(t))^2 + (g'(t))^2}dt. \tag{5.5}$$

Example 5.1

Evaluate the following integrals:

$$\int_C xy^2 \, dx$$

$$\int_C xy^2 \, dy$$

$$\int_C xy^2 \, ds$$

on the quarter circle C defined by $x = 4\cos t$, $y = 4\sin t$, $0 \le t \le \frac{\pi}{2}$ (see Figure 5.3).   ▯

SOLUTION

(a) From (5.3), we have

$$\int_C xy^2 \, dx = \int_0^{\frac{\pi}{2}} (4\cos t)(16\sin^2 t)(-4\sin t \, dt)$$

$$= -256 \int_0^{\frac{\pi}{2}} \sin^3 t \cos t \, dt$$

$$= -256 \left(\frac{1}{4}\sin^4 t \right)_0^{\frac{\pi}{2}} = -64.$$

(b) From (5.4),

$$\int_C xy^2 \, dy = \int_0^{\frac{\pi}{2}} (4\cos t)(16\sin^2 t)(4\cos t \, dt)$$

$$= 256 \int_0^{\frac{\pi}{2}} \sin^2 t \cos^2 t \, dt$$

$$= 256 \int_0^{\frac{\pi}{2}} \frac{1}{4}\sin^2 2t \, dt$$

$$= 64 \int_0^{\frac{\pi}{2}} \frac{1}{2}(1 - \cos 4t) \, dt$$

$$= 32 \left(t - \frac{1}{4}\sin 4t \right)_0^{\frac{\pi}{2}} = 16\pi.$$

(c) From (5.5),

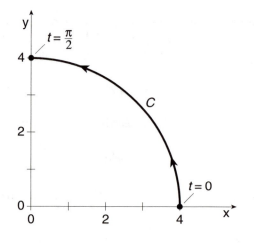

FIGURE 5.3
Definition of quarter-circle

$$\int_C xy^2 ds = \int_0^{\frac{\pi}{2}} (4\cos t)(16\sin^2 t)\sqrt{16(\cos^2 t + \sin^2 t)}dt$$

$$= 256 \int_0^{\frac{\pi}{2}} \sin^2 t \cos t\, dt$$

$$= 256 \left(\frac{1}{3}\sin^3 t\right)_0^{\frac{\pi}{2}} = \frac{256}{3}.$$

∎

5.1.0.2 Method of Evaluation — Curve Defined by an Explicit Function

If the curve C is defined by an explicit function $y = f(x)$, $a \leq x \leq b$, we can use x as a parameter with $dy = f'(x)dx$ and $ds = \sqrt{1 + (f'(x))^2}dx$, the foregoing line integrals become, in turn,

$$\int_C G(x, y)dx = \int_a^b G(x, f(x))dx \qquad (5.6)$$

$$\int_C G(x, y)dy = \int_a^b G(x, f(x))f'(x)dx \qquad (5.7)$$

$$\int_C G(x, y)ds = \int_a^b G(x, f(x))\sqrt{1 + (f'(x))^2}dx. \qquad (5.8)$$

A line integral along a piecewise smooth curve is defined as the sum of the integrals over the various smooth curves whose union comprises C. For example, if C is composed of smooth curves C_1 and C_2, then

$$\int_C G(x,y)ds = \int_{C_1} G(x,y)ds + \int_{C_2} G(x,y)ds.$$

It is common practice to write this sum as one integral without parentheses as

$$\int_C P(x,y)dx + Q(x,y)dy \quad \text{or simply} \quad \int_C Pdx + Qdy. \qquad (5.9)$$

A line integral along a *closed* curve C is very often denoted by

$$\oint_C Pdx + Qdy.$$

Example 5.2
Evaluate $\int_C xydx + x^2dy$, where C is given by $y = x^3$, $-1 \le x \le 2$. ▯

SOLUTION The curve C is illustrated in Figure 5.4 and is defined by the explicit function $y = x^3$. Hence, we can use x as the parameter. Using $dy = 3x^2dx$, we get

$$\int_C xy\,dx + x^2\,dy = \int_{-1}^{2} x(x^3)\,dx + x^2(3x^2dx)$$

$$= \int_{-1}^{2} 4x^4 dx$$

$$= \left(\frac{4}{5}x^5\right)_{-1}^{2} = \frac{132}{5}.$$

Example 5.3
Evaluate $\oint_C xdx$, where C is the circle $x = \cos t$, $y = \sin t$, $0 \le t \le 2\pi$. ▯

SOLUTION From (5.3), we have

$$\oint_C xdx = \int_{0}^{2\pi} \cos t(-\sin tdt) = \left(\frac{1}{2}\cos^2 t\right)_{0}^{2\pi} = \frac{1}{2}(1-1) = 0.$$

FIGURE 5.4
The curve $y = x^3$

Example 5.4
Evaluate $\oint_c y^2 dx - x^2 dy$ on the closed curve C that is shown in Figure 5.5
(a). ☐

SOLUTION   Because C is piecewise smooth, we can express the integral
as a sum of integrals. Symbolically, we write

$$\oint_C = \int_{C_1} + \int_{C_2} + \int_{C_3}$$

where C_1, C_2, and C_3 are the curves shown in Figure 5.5 (b). On C_1, we
use x as a parameter. Because $y = 0$, $dy = 0$; therefore

$$\int_{C_1} y^2 dx - x^2 dy = \int_0^2 (0)dx - x^2(0) = 0.$$

On C_2, we use y as a parameter. From $x = 2$, $dx = 0$, we have

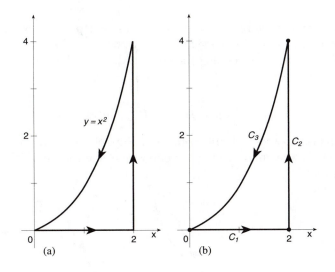

FIGURE 5.5
Integration of closed curve using integral sums

$$\int_{C_2} y^2\, dx - x^2\, dy = \int_0^4 y^2(0) - 4dy$$

$$= -\int_0^4 4dy = (-4y)_0^4 = -16.$$

Finally, on C_3, we again use x as a parameter. From $y = x^2$, we get $dy = 2x\, dx$ and so

$$\int_{C_3} y^2\, dx - x^2\, dy = \int_2^0 x^4 dx - x^2(2x\, dx)$$

$$= \int_2^0 (x^4 - 2x^3)\, dx$$

$$= \left(\frac{1}{5}x^5 - \frac{1}{2}x^4\right)_2^0 = \frac{8}{5}.$$

Therefore,

$$\oint_C y^2\, dx - x^2\, dy = 0 - 16 + \frac{8}{5} = -\frac{72}{5}.$$

It is important to note that a line integral is independent of the parameterization of the curve C provided C is given the same orientation by all sets
of parametric equations defining the curve. Also, recall for definite integrals
that

$$\int_b^a f(x)dx = -\int_a^b f(x)dx.$$

Line integrals possess a similar property. Suppose, as shown in Figure 5.6,
that C denotes the curve having the opposite orientation of C. Then it can
be shown that

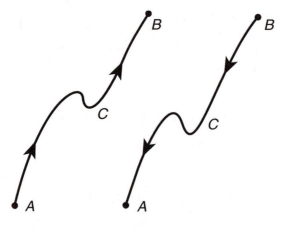

FIGURE 5.6
Line integrals along C and $-C$

$$\int_{-C} Pdx + Qdy = -\int_C Pdx + Qdy$$

or equivalently,

$$\int_{-C} Pdx + Qdy + \int_C Pdx + Qdy = 0. \qquad (5.10)$$

Example 5.5
What is the value of $\int_B^A \frac{dx}{(x+y)}$ along each of the paths shown in Figure 5.7?
[]

SOLUTION   Before this integral can be evaluated, y must be expressed
in terms of x. To do this, recall from the definition of a line integral that
the integrand is always to be evaluated *along the path of integration*.

1. Along the parabolic arc joining A and B, we have $y = x^2$, and making this substitution in the given line integral yields the ordinary definite integral

$$\int_1^2 \frac{dx}{x + x^2} = \int_1^2 \left(\frac{1}{x} - \frac{1}{1+x} \right) dx = \ln(4/3).$$

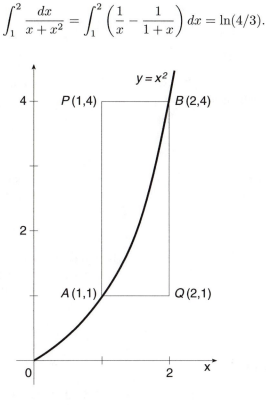

FIGURE 5.7
Possible paths for line integration from $A(1,1)$ to $B(2,4)$

2. Similarly, along the straight-line path from A to B, we have $y = 3x - 2$, and making this substitution in the integrand of the given integral, we obtain the ordinary definite integral

$$\int_1^2 \frac{dx}{x + (3x - 2)} = \frac{1}{4} \ln(4x - 2)\big|_1^2 = \frac{1}{4} (\ln 6 - \ln 2) = \frac{1}{4} \ln(3).$$

3. To compute the line integral along the path APB, we must perform two integrations, one along AP and one along PB, because the relation expressing y in terms of x is different on these two segments. Along AP the integral is obviously zero, because x remains constant and,

therefore, in the sum leading to the integral each Δx_i is zero. Along PB, on which $y = 4$, we have the integral,

$$\int_1^2 \frac{dx}{x+4} = \ln(x+4)|_1^2 = \ln(6/5)$$

that is thus the value of the integral along the entire path APB.

4. Along the path AQB we again have two integrations to perform. Along AQ, on which $y - 1$, we have the integral

$$\int_1^2 \frac{dx}{x+1} = \ln(x+1)|_1^2 = \ln(3/2).$$

Along the vertical segment QB the integral is again zero. Hence, for the entire path AQB the value of the given integral is $\ln(3/2)$.

∎

This example not only illustrates the computational details of line integration, but also shows that, in general, a line integral depends not only on the end points of the integration but also upon the particular path that joins them.

5.2 Line Integrals and Vector Functions

Line integrals of a function G of three variables,

$$\int_C G(x,y,z)dx, \quad \int_C G(x,y,z)dy, \quad \text{and} \quad \int_C G(x,y,z)ds$$

are defined in a manner analogous to Definition 5.1. However, to that list we add a fourth **line integral along a space curve C with respect to z:**

$$\int_C G(x,y,z)dz = \lim_{\|P\|\to 0} \sum_{k=1}^n G(x_k^*, y_k^*, z_k^*)\Delta z_k. \tag{5.11}$$

If C is a smooth curve in 3-space defined by the parametric equations

$$x = f(t) \quad y = g(t) \quad z = h(t) \quad a \le t \le b$$

then the integral in (5.11) can be evaluated by using

$$\int_C G(x,y,z)dz = \int_a^b G(f(t),g(t),h(t))h'(t)dt.$$

The integrals $\int_C G(x,y,z)dz$ and $\int_C G(x,y,z)dy$ are evaluated in a similar manner. The line integral with respect to arc length is

$$\int_C G(x,y,z)ds = \int_a^b G(f(t),g(t),h(t))$$
$$\times \sqrt{(f'(t))^2 + (g'(t))^2 + (h'(t))^2}dt.$$

As in (5.9), in 3-space we are often concerned with line integrals in the form of a sum

$$\int_C P(x,y,z)dx + Q(x,y,z)dy + R(x,y,z)dz.$$

Example 5.6
Evaluate $\int_C ydx + xdy + zdz$, where C is the helix $x = \cos t$, $y = 2\sin t$, $z = t$, $0 \le t \le 2\pi$. $\square$

SOLUTION Substituting the expressions for x, y, and z along with $dx = -2\sin t\,dt$, $dy = 2\cos t\,dt$, $dz = dt$, we obtain

$$\int_C y\,dx + x\,dy + z\,dz = \int_0^{2\pi}(-4\sin^2 t + 4\cos^2 t)\,dt + t\,dt$$
$$= \int_0^{2\pi}(4\cos t + t)\,dt$$
$$= (2\sin 2t + \frac{t^2}{2})\Big|_0^{2\pi} = 2\pi^2.$$

The concept of a vector function of several variables can be used to write a general line integral in a compact fashion. For example, suppose the vector-valued function $\mathbf{F}(x,y) = P(x,y)\mathbf{i} + Q(x,y)\mathbf{j}$ is defined along a curve C : $x = f(t)$, $y = g(t)$, $a \le t \le b$, and suppose $\mathbf{r}(t) = f(t)\mathbf{i} + g(t)\mathbf{j}$ is the position vector of points on C. Then the derivative of $\mathbf{r}(t)$,

$$\frac{d\mathbf{r}}{dt} = f'(t)\mathbf{i} + g'(t)\mathbf{j} = \frac{dx}{dt}\mathbf{i} + \frac{dy}{dt}\mathbf{j}$$

prompts us to define the following:

$$d\mathbf{r} = \frac{d\mathbf{r}}{dt}dt = dx\mathbf{i} + dy\mathbf{j}.$$

Because

$$\mathbf{F}(x, y) \cdot d\mathbf{r} = P(x, y)dx + Q(x, y)dy$$

we can write

$$\int_C P(x, y)dx + Q(x, y)dy = \int_C \mathbf{F} \cdot d\mathbf{r}. \qquad (5.12)$$

Similarly, for a line integral on a space curve,

$$\int_C P(x, y, z)dx + Q(x, y, z)dy + R(x, y, z)dz = \int_C \mathbf{F} \cdot d\mathbf{r} \qquad (5.13)$$

where $\mathbf{F}(x, y, z) = P(x, y, z)\mathbf{i} + Q(x, y, z)\mathbf{j} + R(x, y, z)\mathbf{k}$ and $d\mathbf{r} = dx\mathbf{i} + dy\mathbf{j} + dz\mathbf{k}$.

It may well be asked what physical result is gained by computing a line integral. The answer is: you have a number. Depending on the type of problem, this number may represent work done, change in potential energy, total heat flow, change in entropy, circulation of a fluid and so on. In other words, the physical application determines how a line integral result is interpreted.

5.3   Work

In the beginning courses in calculus or physics it is shown that the work done in moving an object from $x = a$ to $x = b$ by a force $F(x)$, which varies in magnitude but not in direction, is given by the definite integral $W = \int_a^b F(x)\,dx$. From the physical interpretation of the dot product, the work done in moving an object a distance d by a force of magnitude F in the same direction as the force is

$$W = Fd$$

When a constant force $\mathbf{F}$ is applied to a body acting at an angle θ to the direction of motion, then the work done by $\mathbf{F}$ is defined to be

$$W = (|\mathbf{F}| \cos \theta)|\mathbf{d}| = |\mathbf{F}||\mathbf{d}| \cos \theta. \qquad (5.14)$$

This implies that if $\mathbf{F}$ causes a displacement $\mathbf{d}$ of a body, then the work done is

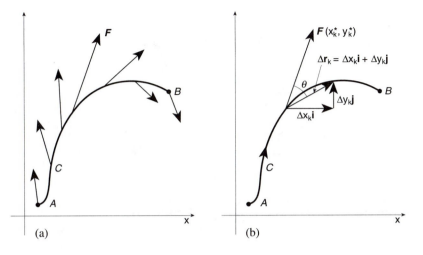

FIGURE 5.8
Force F acting on a smooth curve

$$W = \mathbf{F} \cdot \mathbf{d}. \tag{5.15}$$

In general, a force $\mathbf{F}(x,y) = P(x,y)\mathbf{i} + Q(x,y)\mathbf{j}$ acting at each point on a smooth curve $C : x = f(t)$, $y = g(t)$, $a \leq t \leq b$, varies in both magnitude and direction (see Figure 5.8 (a)).

If A and B are the points $(f(a), g(a))$ and $(f(b), g(b))$, respectively, we need to know what the work done by $\mathbf{F}$ is as its point of application moves along C from A to B. To answer this, suppose C is divided into n sub-arcs of length Δs_k. On each sub-arc $\mathbf{F}(x_k^*, y_k^*)$ is a constant force.

If, as shown in Figure 5.8 (b), the length of the vector $\Delta \mathbf{r}_k = (x_k - x_{k-1})\mathbf{i} + (y_k - y_{k-1})\mathbf{j} = \Delta x_k \mathbf{i} + \Delta y_k \mathbf{j}$ is an approximation to the length of the k-th sub-arc, then the approximate work done by $\mathbf{F}$ over the sub-arc is

$$(|\mathbf{F}(x_k^*, y_k^*)| \cos \theta)|\Delta \mathbf{r}_k| = \mathbf{F}(x_k^*, y_k^*) \cdot \Delta \mathbf{r}_k$$
$$= P(x_k^*, y_k^*)\Delta x_k + Q(x_k^*, y_k^*)\Delta y_k.$$

By summing these elements of work and passing to the limit, we naturally define **work done by F along C** as the line integral

$$W = \int_c P(x,y)dx + Q(x,y)dy \quad \text{or} \quad W = \int_C \mathbf{F} \cdot d\mathbf{r}. \tag{5.16}$$

Equation (5.16) extends to force fields acting at points on a space curve. In this case, work $\int_C \mathbf{F} \cdot dr$ is defined as in (5.13). Now, because

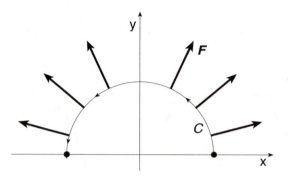

FIGURE 5.9
Half-circle C, with force F perpendicular to C

$$\frac{d\mathbf{r}}{dt} = \frac{d\mathbf{r}}{ds}\frac{ds}{dt}$$

we let $d\mathbf{r} = \mathbf{T}ds$, where $\mathbf{T} = d\mathbf{r}/ds$ is a unit tangent to C. Hence, we have

$$W = \int_C \mathbf{F} \cdot d\mathbf{r} = \int_C \mathbf{F} \cdot \mathbf{T}ds. \tag{5.17}$$

In other words, the work done by a force $\mathbf{F}$ along a curve C is due entirely to the tangential component of $\mathbf{F}$.

Example 5.7
Find the work done by (a) $\mathbf{F} = x\mathbf{i} + y\mathbf{j}$ and (b) $\mathbf{F} = \frac{3}{4}\mathbf{i} + \frac{1}{2}\mathbf{j}$ along the curve C traced by $\mathbf{r}(t) = \cos t\mathbf{i} + \sin t\mathbf{j}$ from $t = 0$ to $t = \pi$. ⬜

SOLUTION

(a) The vector function $\mathbf{r}(t)$ gives the parametric equations $x = \cos t$, $y = \sin t$, $0 \leq t \leq \pi$ which we recognize as a half circle. As seen in Figure 5.9, the force field $\mathbf{F}$ is perpendicular to C at every point. Because the tangential components of $\mathbf{F}$ are zero, the work done along C is zero. To see this we use (5.16):

$$W = \int_C \mathbf{F} \cdot d\mathbf{r} = \int_C (x\mathbf{i} + y\mathbf{j}) \cdot d\mathbf{r}$$

$$= \int_0^\pi (\cos t\mathbf{i} + \sin t\mathbf{j}) \cdot (-\sin t\mathbf{i} + \cos t\mathbf{j})dt$$

$$= \int_0^\pi (-\cos t\sin t + \sin t\cos t)dt = 0.$$

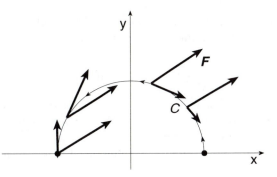

FIGURE 5.10
Projections of F on unit tangent vectors

(b) In Figure 5.10 the vectors tangent to the semi-circle are the projections of **F** on the unit tangent vectors. The work done by **F** is

$$W = \int_c \mathbf{F} \cdot d\mathbf{r} = \int_C \left(\frac{3}{4}\mathbf{i} + \frac{1}{2}\mathbf{j}\right) \cdot d\mathbf{r}$$

$$= \int_0^\pi \left(\frac{3}{4}\mathbf{i} + \frac{1}{2}\mathbf{j}\right) \cdot (-\sin t\mathbf{i} + \cos t\mathbf{j})dt$$

$$= \int_0^\pi \left(-\frac{3}{4}\sin t + \frac{1}{2}\cos t\right) dt$$

$$= \left(\frac{3}{4}\cos t + \frac{1}{2}\sin t\right)\Big|_0^\pi = -\frac{3}{2}.$$

∎

The units of work depend on the units of $|\mathbf{F}|$ and on the units of distance.

5.4   Line Integrals Independent of Path

The following concepts are important to the discussion that follows.
Functions of two variables: The **differential** of a function of two variables $\phi(x, y)$ is

$$d\phi = \frac{\partial \phi}{\partial x}dx + \frac{\partial \phi}{\partial y}dy.$$

A differential expression $P(x,y)dx + Q(x,y)dy$ is said to be an **exact differential** if there exists a function $\phi(x,y)$ such that

$$d\phi = P(x,y)dx + Q(x,y)dy.$$

For example, the expression $x^2y^3\,dx + x^3y^2\,dy$ is an exact differential because it is the differential of $\phi(x,y) = \frac{1}{3}x^3y^3$. On the other hand $(2y^2 - 2y)dx + (2xy - x)dy$ is not an exact differential.

Functions of three variables: The differential of a function of three variables $\phi(x,y,z)$ is

$$d\phi = \frac{\partial\phi}{\partial x}\,dx + \frac{\partial\phi}{\partial y}\,dy + \frac{\partial\phi}{\partial z}\,dz.$$

A differential expression $P(x,y,z)\,dx + Q(x,y,z)\,dy + R(x,y,z)\,dz$ is an exact differential if there exists a function $\phi(x,y,z)$ such that

$$d\phi = P(x,y,z)dx + Q(x,y,z)dy + R(x,y,z)dz.$$

The value of a line integral generally depends on the curve or **path** between two points A and B. There are, however, exceptions. A line integral whose value is the same for *every* curve connecting A and B is said to be **independent of path**.

The following theorem establishes an important relationship between the seemingly disparate notions of path independence and exact differentials. In addition, it provides a means of evaluating path-independent line integrals in a manner analogous to the fundamental theorem of calculus:

$$\int_a^b f'(x)dx = f(b) - f(a).$$

THEOREM 5.1

Suppose there exists a function $\phi(x,y)$ such that $d\phi = Pdx + Qdy$; i.e., $Pdx + Qdy$ is an exact differential. Then $\int_C Pdx + Qdy$ depends on only the endpoints A and B of the path C and

$$\int_C Pdx + Qdy = \phi(B) - \phi(A).$$

PROOF L   et C be a smooth path defined parametrically by $x = f(t)$, $y = g(t)$, $a \le t \le b$, and let the coordinate of A and B be $(f(a), g(a))$ and $(f(b), g(b))$, respectively. Then, by the chain rule,

$$\int_C Pdx + Qdy = \int_a^b \left(\frac{\partial \phi}{\partial x} \frac{dx}{dt} + \frac{\partial \phi}{\partial y} \frac{dy}{dt} \right) dt$$
$$= \int_a^b \frac{d\phi}{dt} dt = (\phi(f(t), g(t)))|_a^b$$
$$= \phi(f(b), g(b)) - \phi(f(a), g(a))$$
$$= \phi(B) - \phi(A).$$

∎

The theorem is also valid for piecewise smooth curves. In addition, the converse of the theorem is also true. Hence, $\int_C Pdx + Qdy$ is independent of path if and only if $Pdx + Qdy$ is an exact differential. We now state, without proof, the following theorem.

THEOREM 5.2
Let P and Q have continuous first partial derivatives in an open, simply connected region. Then $\int_C Pdx + Qdy$ is independent of the path C if and only if

$$\frac{\partial P}{\partial y} = \frac{\partial Q}{\partial x}$$

for all (x, y) in the region.

Example 5.8
Show that the integral $\int_C(x^2 - 2y^3)dx + (x + 5y)dy$ is not independent of the path C. ▯

SOLUTION From $P = x^2 - 2y^3$ and $Q = x + 5y$, we find

$$\frac{\partial P}{\partial y} = -6y^2 \quad \text{and} \quad \frac{\partial Q}{\partial x} = 1.$$

Because $\frac{\partial P}{\partial y} \neq \frac{\partial Q}{\partial x}$ then, by Theorem 5.2, the integral is not independent of the path. In other words, the differential expression $(x^2 - 2y^3)dx + (x + 5y)dy$ is not an exact differential. ∎

Example 5.9
Show that $\int_C(y^2 - 6xy + 6)dx + (2xy - 3x^2)dy$ is independent of any path C between $(-1, 0)$ and $(3, 4)$. Evaluate the integral. ▯

SOLUTION   Identifying $P = y^2 - 6xy + 6$ and $Q = 2xy - 3x^2$ yields

$$\frac{\partial P}{\partial y} = 2y - 6x \quad \text{and} \quad \frac{\partial Q}{\partial x} = 2y - 6x.$$

Because $\frac{\partial P}{\partial y} = \frac{\partial Q}{\partial x}$, the integral is independent of path and so there exists a function ϕ such that

$$\frac{\partial \phi}{\partial x} = y^2 - 6xy + 6 \quad \text{and} \quad \frac{\partial \phi}{\partial y} = 2xy - 3x^2.$$

To find the function ϕ we can integrate either $\frac{\partial \phi}{\partial x}$ or $\frac{\partial \phi}{\partial y}$. Integrating $\frac{\partial \phi}{\partial x}$ with respect to x partially gives

$$\phi = y^2 x - 3x^2 y + 6x + g(y)$$

where $g(y)$ is the "constant" of integration. Taking the partial derivative of the last expression with respect to y and setting the result equal to Q (i.e., $\frac{\partial \phi}{\partial y}$) then gives

$$\frac{\partial \phi}{\partial y} = 2yx - 3x^2 + g'(y) = 2yx - 3x^2$$

that implies $g'(y) = 0$ and so $g(y) = c$, a constant. But because the differentials

$$d(y^2 x - 3x^2 y + 6x + c) \quad \text{and} \quad d(y^2 x - 3x^2 y + 6x)$$

both yield $(y^2 - 6xy + 6)dx + (2xy - 3x^2)dx$, we can drop the constant C and take

$$\phi = xy^2 - 3x^2 y + 6x.$$

It follows from the fundamental theorem for line integrals (Theorem 5.1) that

$$\int_{(-1,0)}^{(3,4)} (y^2 - 6xy + 6)dx + (2xy - 3x^2)dy = \int_{(-1,0)}^{(3,4)} d(xy^2 - 3x^2 y + 6x)$$

$$= \left. (xy^2 - 3x^2 y + 6x) \right|_{(-1,0)}^{(3,4)}$$

$$= (48 - 108 + 18) - (-6)$$

$$= -36.$$

∎

Alternative Solution

Because the integral is independent of the path, we can integrate along any

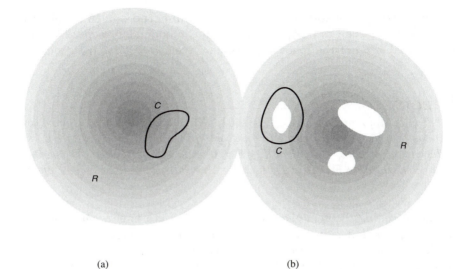

(a)   (b)

FIGURE 5.11
Simple connected regions

convenient curve connecting the given points. In particular, $y = x + 1$ is
such a curve. Using x as a parameter then gives

$$\int_C (y^2 - 6xy + 6)dx + (2xy - 3x^2)dy = \int_{-1}^{3}(-6x^2 - 2x + 7)dx$$
$$= -36.$$

In order to be able to state the next theorem, we need to consider a particular
kind of region in the plane. We say that a region is **simply connected** if:

(a) R is connected, i.e., every pair of points in the region can be joined by
a piecewise smooth curve that lies entirely in R, and

(b) every single closed curve C lying entirely within R can be shrunk, or
contracted to a point without leaving R.

The last statement means that if C is any simple closed curve lying entirely
in R, then the region in the interior C also lies entirely in R. Roughly put,
a simply connected region has no holes in it. In Figure 5.11 (a) we have
illustrated a simply connected region.

The representative simple closed curve C could be shrunk to a point without leaving the region. In Figure 5.11 (b) the region has three distinct holes
in it; because the representative curve C surrounds one of the holes, it could
not be shrunk to a point without leaving the region.

This last region is known as a **multiply connected** one. A simply connected region is said to be **open** if it contains no boundary points. We now state without proof the following theorem giving a test for path independence.

THEOREM 5.3

Let D be a domain in which $P(x,y,z)$, $Q(x,y,z)$ and $R(x,y,z)$ have continuous first derivatives. If the line integral $\int \mathbf{F} \cdot d\mathbf{r} = \int P dx + Q dy + R dz$ is independent of path in D, then $\nabla \times \mathbf{F} = \mathbf{0}$ in D. Conversely, if D is simply connected, the $\nabla \times \mathbf{F} = \mathbf{0}$ in D, then the line integral is independent of path in D.

The preceding theorem gives a simple test to determine whether a given line integral is independent of path: we see whether the curl of $\mathbf{F}$ is zero. Evaluation of a line integral that is independent of path still requires the function $\phi(x,y,z)$, but it is at least nice to know that ϕ exists before searching for it.

Example 5.10

Evaluate $\int_C 2xy e^z dx + (x^2 e^z + y) dy + (x^2 y e^z - z) dz$ along the straight line C from $(0,1,2)$ to $(2,1,-8)$. 　　▯

SOLUTION　　**Method 1:** It is evident that

$$\nabla(x^2 y e^z + \frac{y^2}{2} - \frac{z^2}{2}) = 2xy e^z \mathbf{i} + (x^2 e^z + y)\mathbf{j} + (x^2 y e^z - z)\mathbf{k}$$

and hence the line integral is independent of path. Its value is therefore

$$I = \left(x^2 y e^z + \frac{y^2}{2} - \frac{z^2}{2} \right)_{(0,1,2)}^{(2,1,-8)} = \left(4e^{-8} + \frac{1}{2} - 32 \right) - \left(\frac{1}{2} - 2 \right)$$

$$= 4e^{-8} - 30.$$

Method 2: Because

$$\nabla \times (2xy e^z \mathbf{i} + (x^2 e^z + y)\mathbf{j} + (x^2 y e^z - z)\mathbf{k})$$

$$= \begin{vmatrix} \mathbf{i} & \mathbf{j} & \mathbf{k} \\ \frac{\partial}{\partial x} & \frac{\partial}{\partial y} & \frac{\partial}{\partial z} \\ 2xy e^z & x^2 e^z + y & x^2 y e^z - z \end{vmatrix}$$

$$= (x^2 e^z - x^2 e^z)\mathbf{i} + (2xy e^z - 2xy e^z)\mathbf{j}$$
$$+ (2xe^z - 2xe^z)\mathbf{k} = \mathbf{0}$$

the line integral is independent of path. Thus, there exists a function $\phi(x, y, z)$ such that

$$\nabla\phi = 2xye^z\mathbf{i} + (x^2e^z + y)\mathbf{j} + (x^2ye^z - z)\mathbf{k}$$

or

$$\frac{\partial\phi}{\partial x} = 2xye^z, \quad \frac{\partial\phi}{\partial y} = x^2e^z + y, \quad \frac{\partial\phi}{\partial z} = x^2ye^z - z.$$

Integration of the first of these equations yields

$$\phi(x, y, z) = x^2ye^z + k(y, z).$$

Substitution of this function into the left side of the second equation gives

$$x^2e^z + \frac{\partial k}{\partial y} = x^2e^z + y$$

that implies

$$\frac{\partial k}{\partial y} = y$$

consequently,

$$k(y, z) = \frac{y^2}{2} + L(z)$$

we know both the x- and y-dependence of ϕ

$$\phi(x, y, z) = x^2 + ye^z + \frac{y^2}{2} + L(z).$$

To obtain the z-dependence in $L(z)$, we substitute into the left side of the third equation,

$$x^2ye^z + \frac{dL}{dz} = x^2ye^z - z$$

from which we have

$$\frac{dL}{dz} = -z.$$

Hence, $L(z) = -z^2/2 + c (c$ a constant), and

$$\phi(x, y, z) = x^2ye^z + \frac{y^2}{2} - \frac{z^2}{2} + c.$$

Finally, we have

$$I = \left(x^x y e^z + \frac{y^2}{2} - \frac{z^2}{2}\right)\Bigg|_{(0,1,2)}^{(2,1,-8)} = 4e^{-8} - 30.$$

∎

5.5 Conservative Vector Fields

If $\int_C P\,dx + Q\,dy$ is independent of the path C, we know there exists a function ϕ such that

$$d\phi = \frac{\partial \phi}{\partial x}\,dx + \frac{\partial \phi}{\partial y}\,dy = P\,dx + Q\,dy$$
$$= (P\mathbf{i} + Q\mathbf{j}) \cdot (dx\mathbf{i} + dy\mathbf{j}) = \mathbf{F} \cdot d\mathbf{r}$$

where $\mathbf{F} = P\mathbf{i} + Q\mathbf{j}$ is a vector field and $P = \frac{\partial \phi}{\partial x}$, $Q = \frac{\partial \phi}{\partial y}$. In other words, the vector field $\mathbf{F}$ is a gradient of the function ϕ. Because $\mathbf{F} = \nabla\phi$, $\mathbf{F}$ is said to be a **gradient field** and the function ϕ is said to be a **potential function** for $\mathbf{F}$.

In a gradient force field $\mathbf{F}$, the work done by the force upon a particle moving from position A to position B is the same for all paths between these points. Moreover, the work done by the force along a closed path is zero. For this reason, such a force field is also said to be **conservative**. In a conservative field $\mathbf{F}$ the law of conservation of mechanical energy holds: for a particle moving along a path in a conservative field,

$$\text{kinetic energy} + \text{potential energy} = \text{constant}.$$

In a simply connected region the hypotheses of Theorem 5.3 imply that a force field $\mathbf{F}(x, y) = P(x, y)\mathbf{i} + Q(x, y)\mathbf{j}$ is a gradient(i.e., conservative) if and only if

$$\frac{\partial P}{\partial y} = \frac{\partial Q}{\partial x}.$$

Example 5.11
Show that the vector field $\mathbf{F} = (y^2 + 5)\mathbf{i} + (2xy - 8)\mathbf{j}$ is a gradient field. Find a potential function for $\mathbf{F}$. ⬚

SOLUTION   By identifying $P = y^2 + 5$ and $Q = 2xy - 8$, we see that

$$\frac{\partial P}{\partial y} = \frac{\partial Q}{\partial x} = 2y.$$

Hence, **F** is a gradient field and so there exists a potential function ϕ satisfying

$$\frac{\partial \phi}{\partial x} = y^2 + 5 \quad \text{and} \quad \frac{\partial \phi}{\partial y} = 2xy - 8.$$

Proceeding in the manner of the previous example, we find that $\phi = xy^2 - 8y + 5x$. ∎

Note

$\nabla \phi = \frac{\partial \phi}{\partial x}\mathbf{i} + \frac{\partial \phi}{\partial y}\mathbf{j} = (y^2 + 5)\mathbf{i} + (2xy - 8)\mathbf{j}$.

In general, we can state that in a simply connected region of space, a force field $\mathbf{F}(x, y, z) = P(x, y, z)\mathbf{i} + Q(x, y, z)\mathbf{j} + R(x, y, z)\mathbf{k}$ is conservative if and only if the curl of $\mathbf{F} = \mathbf{0}$ (i.e., $\nabla \times \mathbf{F} = \mathbf{0}$).

5.6 Surface Integrals

Surface Area in the plane the length of an arc of the graph of $y = f(x)$ from $x = a$ to $x = b$ is given by

$$s = \int_a^b \left(\sqrt{1 + \left(\frac{dy}{dx}\right)^2} \right) dx. \tag{5.18}$$

The problem in three dimensions is to find the area $A(s)$ of that portion of the surface S given by a function $z = f(x, y)$ having continuous first partial derivatives on a closed region R in the xy-plane.

Suppose, as shown in Figure 5.12 (a), that an inner partition P of R is formed using lines parallel to the x- and y-axes. P then consists of n rectangular elements R_k of area $\Delta A_k = \Delta x_k \Delta y_k$ that lie entirely within R. Let $(x_k, y_k, 0)$ denote any point in R_k. As we see in Figure 5.12 (a), by projecting the sides of R_k upward, we determine two quantities: a portion S_k of the surface and a portion T_k is a tangent plane at $(x_k, y_k, f(x_k, y_k))$. It seems reasonable to assume that when R_k is small, the area ΔT_k of T_k is approximately the same as the area ΔS_k of S_k.

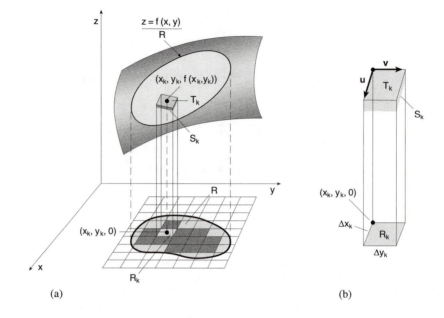

(a)   (b)

FIGURE 5.12
Inner partition P of region R

To find the area of T_k, let us choose $(x_k, y_k, 0)$ at a corner of R_k as shown in Figure 5.12 (b). The indicated vectors **U** and **V**, which form two sides of T_k, are given by

$$\mathbf{U} = \Delta x_k \mathbf{i} + F_x(x_k, y_k)\Delta x_k \mathbf{k}$$
$$\mathbf{V} = \Delta y_k \mathbf{j} + f_y(x_k, y_k)\Delta y_k \mathbf{k}$$

where $f_x(x_k, y_k)$ and $f_y(x_k, y_k)$ are slopes of the lines containing **U** and **V**, respectively. We know that

$$\Delta T_k = |\mathbf{U} \times \mathbf{V}|$$

where

$$\mathbf{U} \times \mathbf{V} = \begin{vmatrix} \mathbf{i} & \mathbf{j} & \mathbf{k} \\ \Delta x_k & 0 & f_x(x_k, y_k)\Delta x_k \\ 0 & \Delta y_k & f_y(x_k, y_k)\Delta y_k \end{vmatrix}$$
$$= (-f_x(x_k, y_k)\mathbf{i} - f_y(x_k, y_k)\mathbf{j} + \mathbf{k})\Delta x_k \Delta y_k.$$

In other words,

$$\Delta T_k = \left(\sqrt{(f_x(x_k, y_k))^2 + (f_y(x_k, y_k))^2 + 1} \right) \Delta x_k \Delta y_k.$$

Consequently, the area A is approximately

$$\sum_{k=1}^{n} \left(\sqrt{1 + (f_x(x_k, y_k))^2 + (f_y(x_k, y_k))^2} \right) \Delta x_k \Delta y_k.$$

Taking the limit of the above sum as $|P| \to 0$ leads us to the following definition.

DEFINITION 5.2   *Let f be a function for which the first partial derivatives f_x and f_y are continuous on a closed region R. Then the* **area of the surface over** *R is given by*

$$A(s) = \int_R \int \left(\sqrt{1 + (f_x(x, y))^2 + (f_y(x, y))^2} \right) dA. \qquad (5.19)$$

Example 5.12
Find the surface area of that portion of the sphere $x^2 + y^2 + z^2 = a^2$ that is above the xy-plane and within the cylinder $x^2 + y^2 = b^2$, $0 < b < a$. ☐

SOLUTION   If we define $z = f(x, y)$ by $f(x, y) = \sqrt{a^2 - x^2 - y^2}$, then

$$f_x(x, y) = \frac{-x}{\sqrt{a^2 - x^2 - y^2}} \quad \text{and} \quad f_y(x, y) = \frac{-y}{\sqrt{a^2 - x^2 - y^2}}$$

and so $1 + (f_x(x, y))^2 + (f_y(x, y))^2 = \frac{a^2}{a^2 - x^2 - y^2}$. Hence, (5.19) is

$$A(s) = \int_R \int \frac{a}{\sqrt{a^2 - x^2 - y^2}} dA$$

where R is indicated in Figure 5.13. To evaluate this double integral, we change to polar coordinates

$$A(s) = a \int_0^{2\pi} \int_0^b (a^2 - r^2)^{-1/2} r \, dr \, d\theta$$

$$= a \int_0^{2\pi} \left(-(a^2 - r^2)^{1/2} \right) \Big|_0^b \, d\theta = a(a - \sqrt{a^2 - b^2}) \int_0^{2\pi} d\theta$$

$$= 2\pi a(a - \sqrt{a^2 - b^2}) \text{ square units.}$$

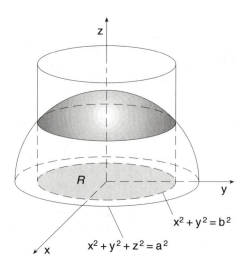

FIGURE 5.13
Integration region R for Example 5.10

As we have seen, double and triple integrals

$$\int_R \int f(x,y)dA \quad \int\int_D \int f(x,y,z)dV$$

are generalizations of the definite integral $\int_a^b f(x)dx$. The surface area integral (5.19) is a generalization of the arc length integral (5.18). We now consider a generalization of the line integral $\int_C G(x,y)ds$. This generalization is called the **surface integral**.

DEFINITION 5.3   *Let G be a function of three variables defined over a region of space containing the surface S. Then the* **surface integral of G over S** *is given by*

$$\int\int_S G(x,y,z)dS = \lim_{\|P\|\to 0}\sum_{k=1}^{n} G(x_k^*, y_k^*, z_k^*)\Delta S_k \qquad (5.20)$$

If G, f, f_x and f_y are continuous throughout a region containing S, then equation (5.20) can be evaluated by means of a double integral. The left side of (5.20) becomes

$$\int\int_S G(x,y,z)dS = \int\int_R G(x,y,f(x,y))$$
$$\times \sqrt{1 + (f_x(x,y))^2 + (f_y(x,y))^2}dA. \quad (5.21)$$

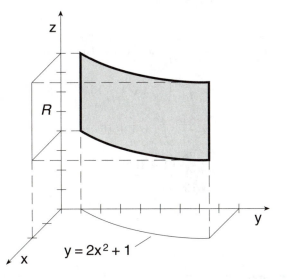

FIGURE 5.14
Rectangular region R in xz-plane

If $y = g(x,z)$ is the equation of a surface S that projects onto a region R of the xz-plane, then

$$\int\int_S G(x,y,z)dS = \int\int_R G(x,g(x,z),z)$$
$$\times \sqrt{1 + (g_x(x,z))^2 + (g_z(x,z))^2}dA. \quad (5.22)$$

Similarly, if $x = h(y,z)$ is the equation of a surface that projects onto the yz-plane, then the analogue of (5.21) is

$$\int\int_S G(x,y,z)dS = \int\int_R G(h(y,z),y,z)$$
$$\times \sqrt{1 + (h_y(y,z))^2 + (h_z(y,z))^2}dA. \quad (5.23)$$

Example 5.13

Evaluate $\int\int_S xz^2\,dS$ where S is that portion of the cylinder $y = 2x^2 + 1$ in the first octant bounded by $x = 0$, $x = 2$, $z = 4$, and $z = 8$. ▯

SOLUTION   We shall use (5.22) with $g(x,z) = 2x^2 + 1$ and R the rectangular region in the xz-plane shown in Figure 5.14 Because $g_x(x,z) = 4x$ and $g_z(x,z) = 0$, it follows that

$$\int\int_S xz^2\,dS = \int_0^2 \int_4^8 xz^2\sqrt{1 + 16x^2}\,dz\,dx$$

$$= \int_0^2 \left(\frac{z^3}{3}x\sqrt{1+16x^2}\right)\Big|_4^8 dx$$

$$= \frac{448}{3}\int_0^2 x(1+16x^2)^{1/2}dx$$

$$= \frac{28}{9}\left(1+16^2\right)^{3/2}\Big|_0^2 = \frac{28}{9}(65^{3/2} - 1)$$

$$= 1627.3.$$

∎

5.7   Orientation of a Surface

Roughly, an orientable surface S, such as that given in Figure 5.15 (a), has two sides that could be painted different colors. The Möbius strip[1] shown in Figure 5.15 (b) is not an orientable surface because it is one-sided. A person who starts to paint the surface of a Möbius strip at a point will paint the entire surface and return to the starting point.

Specifically, we say a smooth surface S is **orientable** or is an **orientable surface** if there exists a continuous unit normal vector function $\mathbf{n}$ defined at each point (x,y,z) on the surface.

The vector field $\mathbf{n}(x,y,z)$ is called the **orientation** of S. But because a unit normal to the surface S at (x,y,z) can be either $\mathbf{n}(x,y,z)$ or $-\mathbf{n}(x,y,z)$, an orientable surface has two orientations (see Figure 5.16 (a)–(c)).

The Möbius strip shown again in Figure 5.15 (d) is not an oriented surface because, if a unit normal $\mathbf{n}$ starts at P on the surface and moves once around

---

[1] August Ferdinand Möbius (1790–1868), German mathematician, student of Gauss, professor of astronomy at Leipzig. He is known for his important work in the theory of surfaces, projective geometry and mechanics. He also contributed to number theory.

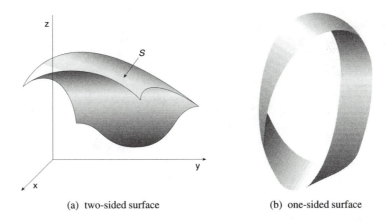

(a) two-sided surface   (b) one-sided surface

FIGURE 5.15
Orientable surfaces

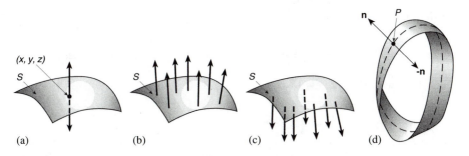

(a)   (b)   (c)   (d)

FIGURE 5.16
Orientable surfaces with unit normals

the strip on the curve C, it ends up on the "opposite side" of the strip at P and so points in the opposite direction.

A surface S defined by $z = f(x, y)$ has an **upward orientation** (Figure 5.16 (b)) when the unit normals are directed upward — i.e., have positive **k** components, and it has a **downward orientation** (Figure 5.16 (c)) when the unit normals are directed downward — i.e., have negative **k** components.

If a smooth surface is defined by $g(x, y, z) = 0$, then a unit normal is

$$\mathbf{n} = \frac{1}{|\nabla g|} \nabla g \qquad (5.24)$$

where $\nabla g = \frac{\partial g}{\partial x}\mathbf{i} + \frac{\partial g}{\partial y}\mathbf{j} + \frac{\partial g}{\partial z}\mathbf{k}$ is the gradient of g. If S is defined by $z = f(x, y)$, then we can use either $g(x, y, z) = z - f(x, y) = 0$ or $g(x, y, z) = f(x, y) - z = 0$ depending on the orientation of S.

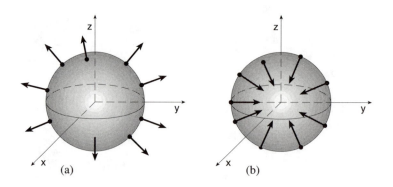

FIGURE 5.17
Sphere with inward and outward n

As an example, consider the sphere of radius $a > 0$: $x^2 + y^2 + z^2 = a^2$. If we define $g(x, y, z) = x^2 + y^2 + z^2 - a^2$, then

$$\nabla g = 2x\mathbf{i} + 2y\mathbf{j} + 2z\mathbf{k} \quad \text{and} \quad |\nabla g| = \sqrt{4x^2 + 4y^2 + 4z^2} = 2a.$$

Then the two orientations of the surface are

$$\mathbf{n} = \frac{x}{a}\mathbf{i} + \frac{y}{a}\mathbf{j} + \frac{z}{a}\mathbf{k} \quad \text{and} \quad \mathbf{n_1} = -\mathbf{n} = -\frac{x}{a}\mathbf{i} - \frac{y}{a}\mathbf{j} - \frac{z}{a}\mathbf{k}.$$

The vector field $\mathbf{n}$ defines an outward orientation, whereas $\mathbf{n_1} = -\mathbf{n}$ defines an inward orientation (see Figure 5.17).

5.8   Volume Integration

The steps leading to the definition of the three-dimensional definite integral or **triple integral** $\int \int \int_D F(x, y, z)dV$ are similar to those of the double integral. A sum of the form $\sum_{k=1}^{n} F(x_k^*, y_k^*, z_k^*)\Delta V_k$, where (x_k^*, y_k^*, z_k^*) is an arbitrary point within each D_k and ΔV_k denotes the volume of each D_k, is called a **Riemman sum**. The type of partition used, where all D_k lie completely within D, is called an **inner partition** of D.

DEFINITION 5.4   *Let F be a function of three variables defined over a closed region D of space. Then the **triple integral** of F over D is given by*

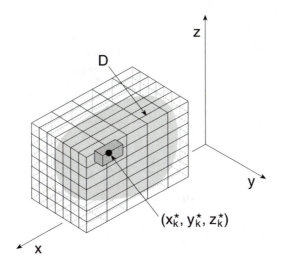

FIGURE 5.18
Inner partition of D

$$\int\int_D\int F(x,y,z)dV = \lim_{|P|\to 0}\sum_{k=1}^{n} F(x_k^*,y_k^*,z_k^*)\Delta V_k \qquad (5.25)$$

If the region D is bounded above by the graph $z = f_2(x,y)$ and bounded below by the graph $z = f_1(x,y)$, it can then be shown that the triple integral in (5.25) can be expressed as a double integral of the partial integral as follows:

$$\int\int_D\int F(x,y,z)dV = \int_R\int\left(\int_{f_1(x,y)}^{f_2(x,y)} F(x,y,z)dz\right)dA$$

where R is the orthogonal projection of D onto the xy-plane. If R is the region shown in Figure 5.19, the triple integral of F over D can be rewritten as the iterated integral

$$\int\int_D\int F(x,y,z)\,dV = \int_{x=a}^{b}\int_{y=g_1(x)}^{g_2(x)}\int_{z=f_1(x,y)}^{f_2(x,y)} F(x,y,z)dz\,dy\,dx. \quad (5.26)$$

To evaluate the integral in (5.26) we begin by evaluating the partial integral

$$\int_{f_1(x,y)}^{f_2(x,y)} F(x,y,z)\,dz$$

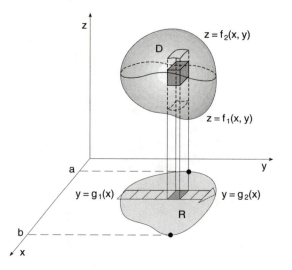

FIGURE 5.19
Type I region R

in which both x and y are held fixed. In a double integral there are only two possible orders of integration, $dy\,dx$ and $dx\,dy$. The triple integral in (5.26) illustrates one of six possible orders of integration:

$dz\,dy\,dx$	$dz\,dx\,dy$	$dy\,dx\,dz$
$dx\,dy\,dz$	$dx\,dz\,dy$	$dy\,dz\,dx$

The last two differentials tell the coordinate plane in which the region R is situated. As an example, the iterated integral corresponding to the order of integration $dx\,dz\,dy$ must have the form

$$\int_{y=c}^{d}\int_{z=k_1(y)}^{k_2(y)}\int_{x=h_1(y,z)}^{h_2(y,z)} F(x,y,z)dx\,dz\,dy.$$

The geometric interpretation of this integral and the region R of integration in the yz-plane are shown in Figure 5.20.
A list of some of the standard applications of the triple integral follows:

- **Mass:**
  If $\rho(x,y,z)$ is density, then the **mass** of the solid D is given by

$$m = \int\int_{D}\int \rho(x,y,z)\,dV.$$

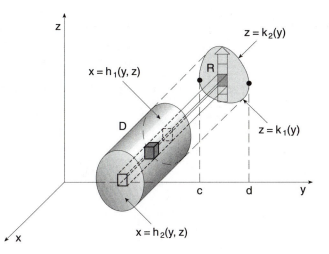

FIGURE 5.20
Geometric interpretation of triple integral

- **First Moments:**
  The **first moments** of the solid about the coordinate planes indicated by the subscripts are given by

$$M_{xy} = \int \int_D \int z\rho(x,y,z)\,dV \quad M_{xz} = \int \int_D \int y\rho(x,y,z)\,dV$$

$$M_{yz} = \int \int_D \int x\rho(x,y,z)\,dV.$$

- **Second Moments:**
  The **second moments**, or **moments of inertia** of D about the coordinate axes indicated by the subscripts, are given by

$$I_x = \int \int_D \int (y^2 + z^2)\rho(x,y,z)\,dV$$

$$I_y = \int \int_D \int (x^2 + z^2)\rho(x,y,z)\,dV$$

$$I_z = \int \int_D \int (x^2 + y^2)\rho(x,y,z)\,dV.$$

Example 5.14

Find the volume of the solid in the first octant bounded by the graphs of $z = 1 - y^2$, $y = 2x$, and $x = 3$.   ⬚

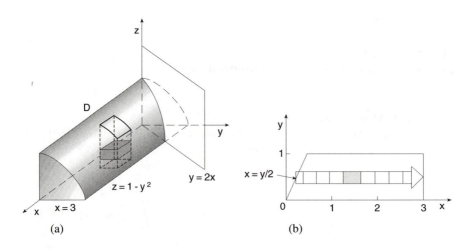

FIGURE 5.21
Integration region in first octant

SOLUTION   As indicated in Figure 5.21 (a), the first integration with respect to z is from 0 to $1 - y^2$. Figure 5.21 (b) shows the projection of the solid D in the xy-plane. Hence, we next integrate with respect to x, from $y/2$ to 3. The last integration is with respect to y from 0 to 1. Thus, we have

$$V = \int \int \int_{D} dV = \int_{0}^{1} \int_{y/2}^{3} \int_{0}^{1-y^{2}} dz dx dy$$

$$= \int_{0}^{1} \int_{y/2}^{3} (1 - y^{2}) dx dy$$

$$= \int_{0}^{1} (x - xy^{2})\big|_{y/2}^{3} dy$$

$$= \int_{0}^{1} \left(3 - 3y^{3} - \frac{1}{2}y + \frac{1}{2}y^{3} \right) dy$$

$$= \left(3y - y^{3} - \frac{1}{4}y^{2} + \frac{1}{8}y^{4} \right)\bigg|_{0}^{1}$$

$$= \frac{15}{8} \quad \text{cubic units.}$$

∎

Example 5.15
Find the volume in the first octant cut from the cylinder $x^2 + z^2 = 4$ by the plane $y + z = 6$.   ▯

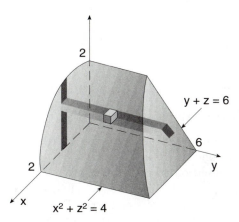

FIGURE 5.22
Region of integration in first octant

SOLUTION   From the figure above, we have

$$V = \int\int\int_D dV = \int_0^2 \int_0^{\sqrt{4-x^2}} \int_0^{6-z} dy\,dz\,dx$$

$$= \int_0^2 \int_0^{\sqrt{4-x^2}} (6-z)dz\,dx$$

$$= \int_0^2 \left(6z - \frac{z^2}{2}\right)\Big|_0^{\sqrt{4-x^2}} dx$$

$$= \int_0^2 \left(\sqrt{4-x^2} - \frac{1}{2}(4-x^2)\right) dx.$$

Polar coordinates can now be used in the first term of the preceding integral by setting $x = 2\sin\theta$ so that $dx = 2\cos\theta d\theta$, which leads to

$$V = 6\int_0^{\pi/2} (2\cos\theta)2\cos\theta d\theta + \left(-2x + \frac{x^3}{6}\right)_0^2$$

$$= 12\int_0^{\pi/2} (1+\cos\theta)d\theta - \frac{8}{3}$$

$$= 12\left(\theta + \frac{1}{2}\sin\theta\right)\Big|_0^{\pi/2} - \frac{8}{3} = 6\pi - \frac{8}{3} \quad \text{cubic units.}$$

∎

5.9   Triple Integrals in Cylindrical Coordinates

When evaluating double integrals, polar coordinates are often used for convenience and simplicity. In a similar manner, other coordinate systems can simplify the evaluation of triple integrals. Two of the most common are cylindrical and spherical coordinates. Cylindrical coordinates are useful in problems involving an axis of symmetry. They are based on a cartesian coordinate along the axis of symmetry and polar coordinates in a plane perpendicular to the axis of symmetry. If the z-axis is the axis of symmetry and polar coordinates are defined in the xy-plane with the origin as pole and the positive x-axis as polar axis, then cylindrical coordinates and cartesian coordinates are related by the equations (see Figure 5.23)

$$x = r\cos\theta, \quad y = r\sin\theta, \quad z = z. \tag{5.27}$$

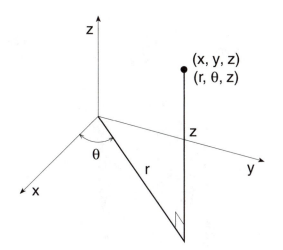

FIGURE 5.23
Cylindrical coordinate system

Recall that r can be expressed in terms of x and y by

$$r = \sqrt{x^2 + y^2}$$

and θ is defined implicitly by the equations

$$\cos \theta = \frac{x}{\sqrt{x^2 + y^2}}, \quad \sin \theta = \frac{y}{\sqrt{x^2 + y^2}}.$$

To use cylindrical coordinates in the evaluation of triple integrals, we must express equations of surfaces in terms of these coordinates. This is very simple. If $F(x, y, z) = 0$ is the equation of a surface in cartesian coordinates, then to express this equation in cylindrical coordinates we substitute from equation (5.27):

$$F(r \cos \theta, r \sin \theta, z) = 0.$$

For example, the right circular cylinder $x^2 + y^2 = 9$, which has the z-axis as its axis of symmetry, has the very simple equation $r = 3$ in cylindrical coordinates.

Suppose we are to evaluate the triple integral of a continuous function $F(x, y, z)$ over some region V of space. The choice of triple iterated integral in cylindrical coordinates implies a subdivision of V into small volumes dV by means of coordinate surfaces $r = $ constant, $\theta = $ constant, and $z = $ constant (Figure 5.24).

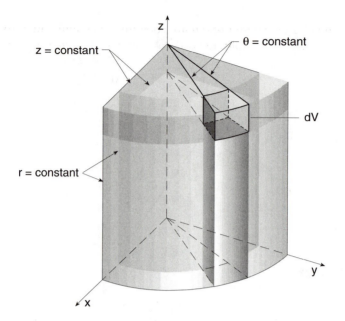

FIGURE 5.24
First quadrant portion of cylinder

Surfaces r = constant are right circular cylinders coaxial with the z-axis; surfaces θ = constant are planes containing the z-axis and therefore perpendicular to the xy-plane; surfaces z = constant are planes parallel to the xy-plane. If we denote small variations in r, θ and z for the element dV by dr, $d\theta$, and dz (see Figure 5.25), then the volume of the element is approximately $(r\,dr\,d\theta)dz$, where $r\,dr\,d\theta$ is the polar cross-sectional area parallel to the xy-plane. Hence, in cylindrical coordinates, we get

$$dV = r\,dz\,dr\,d\theta. \qquad (5.28)$$

The integrand $F(x, y, z)$ is expressed in cylindrical coordinates as

$$F(r\cos\theta, r\sin\theta, z).$$

It remains only to affix appropriate limits to the triple integral. The most commonly used triple iterated integral is with respect to z, r, and θ and, in this case, the z-integration adds the quantities

$$F(r\cos\theta, r\sin\theta, z)r\,dz\,dr\,d\theta$$

in a vertical column, where r and θ are constant. The limits identify surfaces in which each and every column starts and stops and generally depend on r and θ:

$$\int_{h_1(r,\theta)}^{h_2(r,\theta)} F(r\cos\theta, r\sin\theta, z)r\,dz\,dr\,d\theta.$$

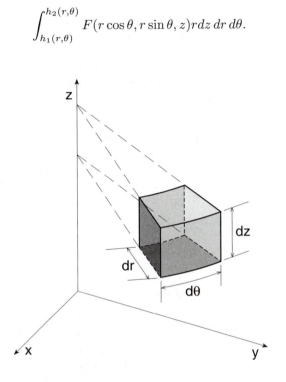

FIGURE 5.25
Volume element

The remaining integrations with respect to r and θ perform additions over the area in the xy-plane onto which all vertical columns project. Because r and θ are simply polar coordinates, the r-integration adds over small areas in a wedge and the θ-integration adds over all wedges. The triple integral with respect to z, r and θ has the form

$$\int_{\theta=a}^{b} \int_{r=g_1(\theta)}^{g_2(\theta)} \int_{z=h_1(\theta)}^{h_2(\theta)} F(r\cos\theta, r\sin\theta, z)r\,dz\,dr\,d\theta.$$

Example 5.16
Find the volume inside both the sphere $x^2 + y^2 + z^2 = 2$ and the cylinder $x^2 + y^2 = 1$. ☐

SOLUTION   The required volume is eight times the first octant volume shown in Figure 5.26. If we use cylindrical coordinates, the volume of an elemental piece is $r\,dz\,dr\,d\theta$.

A z-integration adds these pieces to give volume in a vertical column:

$$\int_{z=0}^{\sqrt{2-r^2}} r\,dz\,dr\,d\theta$$

where limits indicate that for volume in the first octant all columns start on the xy-plane (where $z=0$) and end on the sphere (where $r^2 + z^2 = 2$). An r-integration now adds volumes of all columns that stand on a wedge:

$$\int_{r=0}^{1}\int_{z=0}^{\sqrt{2-r^2}} r\,dz\,dr\,d\theta$$

where limits indicate that all wedges start at the origin (where $r=0$) and end on the curve $x^2 + y^2 = 1$ (or $r = 1$) in the xy-plane. This integration yields the volume of a slice (Figure 5.26). Finally, θ-integration adds volumes of all such slices

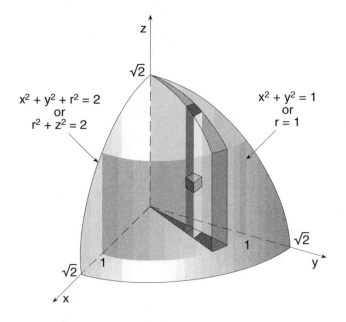

FIGURE 5.26
First octant volume element

$$\int_{\theta=0}^{\pi/2}\int_{r=0}^{1}\int_{z=0}^{\sqrt{2-r^2}} r\,dz\,dr\,d\theta$$

where limits 0 and $\pi/2$ identify positions of first and last wedges, respectively, in the first quadrant. The required volume is then

$$8\int_{\theta=0}^{\pi/2}\int_{r=0}^{1}\int_{z=0}^{\sqrt{2-r^2}} rdz\,dr\,d\theta = 8\int_{0}^{\pi/2}\int_{0}^{1} r\sqrt{2-r^2}dr\,d\theta$$

$$= \int_{0}^{\pi/2}\left(-\frac{1}{3}(2-r^2)^{3/2}\right)_{0}^{1}d\theta$$

$$= \frac{8}{3}(2\sqrt{2}-1)\int_{0}^{\pi/2}d\theta = \frac{4\pi}{3}(2\sqrt{2}-1).$$

∎

5.10 Triple Integrals in Spherical Coordinates

Spherical coordinates are useful in solving problems where figures are symmetric about a point. If the origin is that point, then spherical coordinates (ρ, θ, ϕ) in Figure 5.27

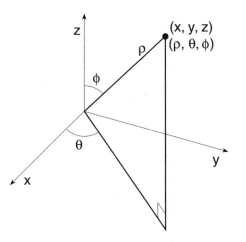

FIGURE 5.27
Spherical coordinate system

are related to cartesian coordinates (x, y, z) by the equations

$$x = \rho \sin \phi \cos \theta$$
$$y = \rho \sin \phi \sin \theta \qquad \qquad (5.29)$$
$$z = \rho \cos \phi.$$

As is the case for polar and cylindrical coordinates, without restrictions on ρ, θ and ϕ, each point in space has many sets of spherical coordinates. The positive value of its spherical coordinate ρ is given by

$$\rho = \sqrt{x^2 + y^2 + z^2}.$$

The θ-coordinate in cylindrical and spherical coordinates is identical so no simple formula for θ in terms of x, y, and z exists. The value of θ lies in the range $0 \leq \theta \leq 2\pi$.

The ϕ-coordinate is the angle between the positive z-axis and the line joining the origin to the point (x, y, z), and that value of ϕ in the range $0 \leq \phi \leq \pi$ is determined by the formula

$$\phi = \cos^{-1} \left(\frac{z}{\sqrt{x^2 + y^2 + z^2}} \right).$$

To transform equations of surfaces, $F(x, y, z) = 0$, from cartesian to spherical coordinates, we substitute from equations (5.29):

$$F(\rho \sin \phi \cos \theta, \rho \sin \phi \sin \theta, \rho \cos \phi) = 0.$$

For example, the sphere $x^2 + y^2 + z^2 = 4$ is symmetric about its center, and its equation in spherical coordinates is simply $\rho = 2$.

The choice of a triple iterated integral in spherical coordinates implies a subdivision of V into small volumes by means of coordinate surfaces $\rho = $ constant, $\theta = $ constant, and $\phi = $ constant (see Figure 5.28). Surfaces $\rho = $ constant are spheres centered at the origin; surfaces $\theta = $ constant are planes containing the z-axis; and surfaces $\phi = $ constant are right circular cones symmetric about the z-axis with the origin as apex.

If we denote small variations in ρ, θ, and ϕ for the element dV by $d\rho$, $d\theta$, and $d\phi$ (Figure 5.29) and approximate dV by a rectangular parallelepiped with dimensions $d\rho$, $\rho d\phi$, and $\rho \sin \phi d\theta$, then

$$dV = (\rho \sin \phi d\theta)(\rho d\phi)d\rho = \rho^2 \sin \phi d\rho \, d\theta \, d\phi. \qquad (5.30)$$

The integrand $F(x, y, z)$ is expressed in spherical coordinates as

$$F(\rho \sin \phi \cos \theta, \rho \sin \phi \sin \theta, \rho \cos \phi).$$

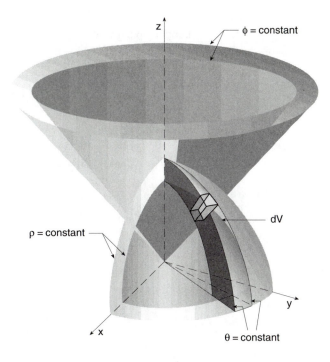

FIGURE 5.28
Triple integration using spherical coordinates

Affixing the appropriate limits is the only remaining duty. The limits depend on which of the six possible triple iterated integrals in spherical coordinates we choose. If we use a triple iterated integral with respect to ρ, ϕ, and θ, then its form is

$$\int\int_V\int F(x,y,z)dV = \int_{\theta=a}^{b}\int_{\phi=g_1(\theta)}^{g_2(\theta)}\int_{\rho=h_1(\theta,\phi)}^{h_2(\theta,\phi)} F(\rho\sin\phi\cos\theta,$$
$$\rho\sin\phi\sin\theta, \rho\cos\phi)\rho^2\sin\phi d\rho\,d\phi\,d\theta. \qquad (5.31)$$

Example 5.17
Find the volume of a sphere.   ▯

SOLUTION   The equation of a sphere of radius R centered at the origin is $x^2 + y^2 + z^2 = R^2$ or, in spherical coordinates, $\rho = R$. The volume of this sphere is eight times the first octant volume shown in Figure 5.30. If we use

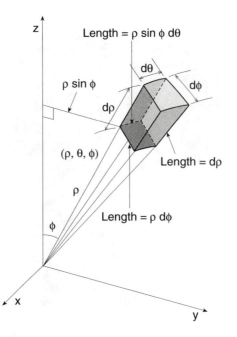

FIGURE 5.29
Spherical coordinate volume element

spherical coordinates, the volume of an elemental piece is

$$\rho^2 \sin \phi d\rho \, d\phi \, d\theta.$$

A ρ-integration adds these volumes for constant ϕ and θ to give the volume in a "spike"

$$\int_0^R \rho^2 \sin \phi d\rho \, d\phi \, d\theta$$

where limits indicate that all spikes start at the origin (where $\rho = 0$) and end on the sphere (where $\rho = R$). A ϕ-integration now adds the volumes of spikes for constant θ. This yields the volume of a slice

$$\int_0^{\pi/2} \int_0^R \rho^2 \sin \phi d\rho \, d\phi \, d\theta$$

where limits indicate that all slices in the first octant start on the z-axis (where $\phi = 0$) and end on the xy-plane (where $\phi = \frac{\pi}{2}$). Finally, the θ-integration adds the volumes of all such slices

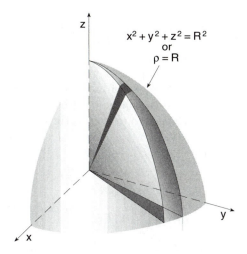

FIGURE 5.30
First octant volume element for Example 5.17

$$\int_0^{\pi/2} \int_0^{\pi/2} \int_0^R \rho^2 \sin\phi d\rho \, d\phi \, d\theta$$

where limits 0 and $\pi/2$ identify positions of first and last slices, respectively, in the first octant. The volume of the sphere is given by

$$\begin{aligned}
8\int_0^{\pi/2} \int_0^{\pi/2} \int_0^R \rho \sin\phi d\rho \, d\phi \, d\theta &= 8\int_0^{\pi/2} \int_0^{\pi/2} \left(\frac{\rho^3}{3}\sin\phi\right)_0^R d\phi \, d\theta \\
&= \frac{8R^3}{3}\int_0^{\pi/2} (-\cos\phi)_0^{\pi/2} \, d\theta \\
&= \frac{8R^3}{3}\,(\theta)_0^{\pi/2} = \frac{4\pi R^3}{3}.
\end{aligned}$$

∎

Example 5.18
Find the center of mass of a solid object of constant density if it is in the shape of a right circular cone.   ⬚

SOLUTION   Let the cone have altitude h and base radius R. Then its mass is

$$M = \frac{1}{3}\pi R^2 h\,\rho,$$

where ρ is the density of the object. If we place axes as shown in Figure 5.31, then $\bar{x} = \bar{y} = 0$; i.e., the center of mass is on the axis of symmetry of the cone. To find $\bar{z}$ we offer three solution methods.

Method 1: If we use cartesian coordinates, the equation of the surface of the cone is of the form $z = k\sqrt{x^2 + y^2}$. Because $(0, R, h)$ is a point on the cone, $h = kR$, and therefore $k = H/R$. Now

$$M\bar{z} = 4\int_0^R \int_0^{\sqrt{R^2-x^2}} \int_{k\sqrt{x^2+y^2}}^h z\rho\,dz\,dy\,dx$$

$$= 4\rho\int_0^R \int_0^{\sqrt{R^2-x^2}} \left\{\frac{z^2}{2}\right\}_{k\sqrt{x^2+y^2}}^h dy\,dx$$

$$= 2\rho\int_0^R \int_0^{\sqrt{R^2-x^2}} \left\{h^2 - k^2(x^2+y^2)\right\} dy\,dx$$

$$= 2\rho\int_0^R \left\{h^2 y - k^2 x^2 y - \frac{k^2 y^3}{3}\right\}_0^{\sqrt{R^2-x^2}} dx$$

$$= 2\rho\int_0^R \left\{h^2\sqrt{R^2-x^2} - k^2 x^2\sqrt{R^2-x^2}\right.$$
$$\left. - \frac{k^2}{3}(R^2-x^2)^{3/2}\right\} dx.$$

If we now set $x = R\sin\theta$, then $dx = R\cos\theta\,d\theta$, and

$$M\bar{z} = 2\rho\int_0^{\pi/2} \left\{h^2 R\cos\theta - k^2(R^2\sin^2\theta)R\cos\theta - \frac{k^2}{3}\cos^3\theta\right\} R\cos\theta\,d\theta$$

$$= 2\rho R^2\int_0^{\pi/2} \left\{\frac{h^2}{2}(1+\cos 2\theta) - \frac{k^2 R^2}{8}(1-\cos 4\theta)\right.$$
$$\left. - \frac{k^2 R^2}{12}\left(1+2\cos 2\theta + \frac{1+\cos 4\theta}{2}\right)\right\} d\theta$$

$$= \rho R^2\left\{h^2\left(\theta + \frac{\sin 2\theta}{2}\right) - \frac{k^2 R^2}{4}\left(\theta - \frac{\sin 4\theta}{4}\right)\right.$$
$$\left. - \frac{k^2 R^2}{6}\left(\frac{3\theta}{2} + \sin 2\theta + \frac{\sin 4\theta}{8}\right)\right\}_0^{\pi/2}$$

$$= \rho R^2\left\{\frac{\pi h^2}{2} - \frac{k^2 R^2\pi}{8} - \frac{k^2 R^2\pi}{8}\right\} = \frac{\pi R^2 h^2\rho}{4}.$$

Therefore,

$$\bar{z} = \frac{\pi R^2 h^2 \rho}{4} \frac{3}{\pi R^2 h \rho} = \frac{3h}{4}.$$

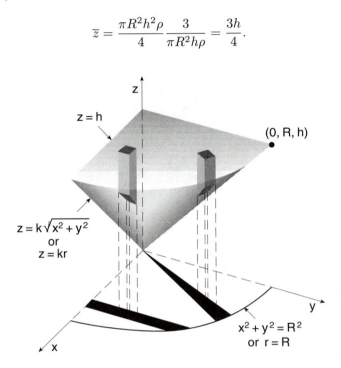

FIGURE 5.31
First quadrant element for Example 5.18

Method 2: If cylindrical coordinates are used, the equation of the surface of the cone is $z = kr$. From Figure 5.31, we see that

$$M\bar{z} = 4 \int_0^{\pi/2} \int_0^R \int_{kr}^h z\rho r\, dz\, dr\, d\theta = 4\rho \int_0^{\pi/2} \int_0^R \left\{ \frac{rz^2}{2} \right\}_{kr}^h dr\, d\theta$$

$$= 2\rho \int_0^{\pi/2} \int_0^R r(h^2 - k^2 r^2)\, dr\, d\theta = 2\rho \int_0^{\pi/2} \left\{ \frac{r^2 h^2}{2} - \frac{k^2 r^4}{4} \right\}_0^R d\theta$$

$$= \rho \left(R^2 h^2 - \frac{k^2 R^4}{2} \right) \{\theta\}_0^{\pi/2} = \rho \left(R^2 h^2 - \frac{k^2 R^4}{2} \right) \frac{\pi}{2} = \frac{\pi R^2 h^2 \rho}{4}.$$

Here again, we have

$$\bar{z} = 3h/4.$$

Method 3: In using spherical coordinates, we take $\rho = \gamma$ to avoid confusion with the density. The equation of the surface of the cone then becomes

$$\phi = \cos^{-1}\left(\frac{h}{\sqrt{h^2 + R^2}}\right) = \phi_1.$$

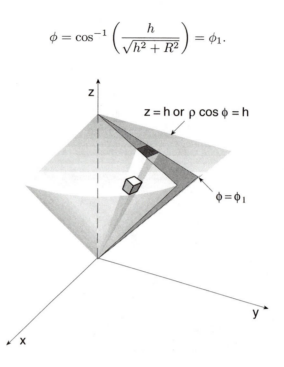

FIGURE 5.32
Spherical coordinate volume element for Example 5.18

From Figure 5.32, we see that

$$
\begin{aligned}
M\bar{z} &= 4\int_0^{\pi/2}\int_0^{\phi_1}\int_0^{h\sec\phi}(\gamma\cos\phi)\rho\,\gamma^2\sin\phi\,d\gamma\,d\phi\,d\theta \\
&= 4\rho\int_0^{\pi/2}\int_0^{\phi_1}\left\{\frac{\gamma^4}{4}\sin\phi\cos\phi\right\}_0^{h\sec\phi}d\phi\,d\theta \\
&= \rho h^4\int_0^{\pi/2}\int_0^{\phi_1}\left(\frac{\sin\phi}{\cos^3\phi}\right)d\phi\,d\theta = \rho h^4\int_0^{\pi/2}\left\{\frac{1}{2\cos^2\phi}\right\}_0^{\phi_1}d\theta \\
&= \frac{\rho h^4}{2}\left\{\frac{1}{\cos^2\phi}-1\right\}\{\theta\}_0^{\pi/2} \\
&= \frac{\pi\rho h^4}{4}\left\{\frac{h^2+R^2}{h^2}-1\right\} = \frac{\pi h^2 R^2\rho}{4}.
\end{aligned}
$$

Once again, $\bar{z} = 3h/4$ as before.

Exercises

1. Evaluate the line integral, where C is the given curve.

    (a) $\int_C x ds$, $C: x = t^3, y = t, \ 0 \le t \le 1$

    (b) $\int_C xy^4 ds$, C is the right half of the circle $x^2 + y^2 = 16$

    (c) $\int_C (x - 2y^2) dy$, C is the arc of the parabola $y = x^2$ from $(-2, 4)$ to $(2, 3)$

    (d) $\int_C xy dx + (x - y) dy$, C consists of line segments from $(0, 0)$ to $(2, 0)$ and from $(2, 0)$ to $(3, 2)$

    (e) $\int_C x\sqrt{y} dx + 2y\sqrt{x} dy$, C consists of the arc of the circle $x^2 + y^2 = 1$ from $(1, 0)$ to $(0, 1)$ and the line segment from $(0, 1)$ to $(4, 3)$

2. Evaluate the line integral in the following questions.

    (a) $\int_C xz ds$ along the first octant part of $y = x^2$, $z + y = 1$ from $(0, 0, 1)$ to $(1, 1, 0)$

    (b) $\int_C x\sqrt{y + z} ds$ where C is that part of the curve $3x + 2y + 3z = 6$, $x - 2y + 4z = 5$ from $(1, 0, 1)$ to $(0, 9/14, 11/7)$

    (c) $\int_C (x + y) z ds$ where C is the curve $y = x$, $z = 1 + y^4$ from $(-1, -1, 2)$ to $(1, 1, 2)$

    (d) $\int_C (2y + yz) ds$ where C is the curve $z = xy$, $x = y^2$ from $(0, 0, 0)$ to $(4, -2, -8)$

3. Find the work done by a force $\mathbf{F} = x\mathbf{i} + y\mathbf{j}$ on a particle as it moves once counter clockwise around the ellipse $b^2 x^2 + a^2 y^2 = a^2 b^2$, $z = 0$.

4. Evaluate the line integral

$$\int_C \frac{x^3}{(1 - x^4)^3} dx + y^2 e^y dy + \frac{z}{\sqrt{1 + z^2}} dz$$

    where C consists of line segments joining successively the points $(0, -1, 1)$, $(1, -1, 1)$, $(1, 0, 1)$ and $(1, 0, 2)$

5. The cycloid $x = R(\theta - \sin \theta)$, $y = R(1 - \cos \theta)$ is the curve traced out by a fixed point on the circumference of a circle of radius R rolling along the x-axis. Suppose the point is acted on by a force of unit magnitude directed toward the center of the rolling circle.

    (a) Find the work done by the force as the point from $\theta = 0$ to $\theta = \pi$.

    (b) How much of the work in (a) is done by the vertical component of the force?

6. Evaluate the line integral $\int_C \mathbf{F} \cdot d\mathbf{r}$, where C is given by the vector function $\mathbf{r}(t)$.

(a) $\mathbf{F}(x,y) = x^2 y\mathbf{i} - xy\mathbf{j}$, $\mathbf{r}(t) = t^3\mathbf{i} + t^4\mathbf{j}$, $0 \le t \le 1$

(b) $\mathbf{F}(x,y,z) = (y+z)\mathbf{i} - x^2\mathbf{j} - 4y^2\mathbf{k}$ $\mathbf{r}(t) = t\mathbf{i} + t^2\mathbf{j} + t^4\mathbf{k}$, $0 \le t \le 1$

(c) $\mathbf{F}(x,y,z) = \sin x\mathbf{i} + \cos y\mathbf{j} + xz\mathbf{k}$, $\mathbf{r}(t) = t^3\mathbf{i} - t^2\mathbf{j} + t\mathbf{k}$, $0 \le t \le 1$

(d) $\mathbf{F}(x,y,z) = x^2\mathbf{i} + xy\mathbf{j} + z^2\mathbf{k}$, $\mathbf{r}(t) = \sin t\mathbf{i} + \cos t\mathbf{j} + t^2\mathbf{k}$, $0 \le t \le \pi/2$

7. Evaluate the line integral $\int_C \mathbf{F} \cdot d\mathbf{r}$ where $\mathbf{F}(x,y) = e^{x-1}\mathbf{i} + xy\mathbf{j}$ and C is given by $\mathbf{r}(t) = t^2\mathbf{i} + t^3\mathbf{j}$, $0 \le t \le 1$.

8. Find the center of mass of a wire in the shape of the helix $x = 2\sin t$, $y = 2\cos t$ $z = 3t$, $0 \le t \le 2\pi$ if the density is a constant k.

9. Find the mass and center of mass of a wire in the shape of the helix $x = t$, $y = \cos t$, $z = \sin t$, $0 \le t \le 2\pi$, if the density at any point is equal to the square of the distance from the origin.

10. Determine whether or not $\mathbf{F}$ is a conservative field. If it is, find a function f such that $\mathbf{F} = \nabla f$.

(a) $\mathbf{F}(x,y) = (3x^2 - 4y)\mathbf{i} + (4y^2 - 2x)\mathbf{j}$

(b) $\mathbf{F}(x,y) = (x^2 + y)\mathbf{i} + (y^2 + x)\mathbf{j}$

(c) $\mathbf{F}(x,y) = (1 + 4x^3y^3)\mathbf{i} + 3x^4y^2\mathbf{j}$

(d) $\mathbf{F}(x,y) = (y\cos x - \cos y)\mathbf{i} + (\sin x + x\sin y)\mathbf{j}$

(e) $\mathbf{F}(x,y) = (ye^x + \sin y)\mathbf{i} + (e^x + x\cos y)\mathbf{j}$

In the next two questions show that the line integral is independent of path and evaluate the integral.

11. $\int_C 2x\sin y\,dx + (x^2\cos y - 3y^2)dy$ C is any path from $(-1,0)$ to $(5,1)$.

12. $\int_C (2y^2 - 12x^3y^3)dx + (4xy - 9x^4y^2)dy$, C is any path from $(1,1)$ to $(3,2)$.

13. Let $\mathbf{F}(x,y) = \frac{-y\mathbf{i}+x\mathbf{j}}{x^2+y^2}$

(a) Show that $\frac{\partial P}{\partial y} = \frac{\partial Q}{\partial x}$.

(b) Show that $\int_C \mathbf{F} \cdot d\mathbf{r}$ is not independent of path. [*Hint:* Show that $\int_{C_1} \mathbf{F} \cdot d\mathbf{r}$ and $\int_{C_2} \mathbf{F} \cdot d\mathbf{r}$, where the regions C_1 and C_2 are the upper and lower halves of the circle $x^2 + y^2 = 1$ from $(1,0)$ to $(-1,0)$]. Does this contradict Theorem 5.3?

14. (a) Suppose that $\mathbf{F}$ is an inverse square force field, i.e.,

$$\mathbf{F}(\mathbf{r}) = \frac{c\mathbf{r}}{|\mathbf{r}|^3}$$

for some constant c, where $\mathbf{r} = x\mathbf{i} + y\mathbf{j} + z\mathbf{k}$. Find the work done by $\mathbf{F}$ in moving an object from a point P_1 along a path to a point P_2 in terms of the distances d_1 and d_2 from these points to the origin.

(b) An example of an inverse square field is the gravitational field $\mathbf{F} = -(mMG)/|\mathbf{r}|^3$. Use part (a) to find the work done by the gravitational field when the earth moves from aphelion (at a maximum distance of $1.52 \times 10^8\ km$ from the sun) to perihelion (at a minimum distance of $1.47 \times 10^8\ km$). Use the values $m = 5.97 \times 10^{24}\ kg$, $M = 1.99 \times 10^{30}\ kg$, and $G = 6.67 \times 10^{-11}\ Nm^2/kg^2$.

(c) Another example of an inverse square field is the electrical field $\mathbf{E} = \varepsilon qQ\mathbf{r}/|\mathbf{r}|^3$. Suppose an electron with a charge of $-1.6 \times 10^{-19}\ C$ is at the origin. A positive charge is at a distance $10^{-12}\ m$ from the electron. Use part (a) to find the work done by the electrical field. (Use the value $\varepsilon = 8.985 \times 10^{10}$.)

15. Evaluate the given surface integral:

(a) $\int\int_S y\,ds$, S is the part of the plane $3x + 2y + z = 6$ that lies on the first octant.

(b) $\int\int_S xz\,ds$, S is the triangle with vertices $(1,0,0)$, $(0,1,0)$ and $(0,0,1)$.

(c) $\int\int_S x\,ds$, S is the surface $y = x^2 + 4z$, $0 \le x \le 2$, $0 \le z \le 2$.

(d) $\int\int_S (y^2 + z^2)\,ds$, S the part of the paraboloid $x = 4 - y^2 - z^2$ that lies in front of the plane $x = 0$.

16. A fluid with density 1200 flows with velocity $\mathbf{V} = y\mathbf{i} + x\mathbf{j} + 2z\mathbf{k}$. Find the rate of flow upward through the paraboloid $z = 9 - (x^2 + y^2)/4$, $x^2 + y^2 \le 36$.

17. Evaluate the given surface integral:

(a) $\int\int_S (yz^2\mathbf{i} + ye^x\mathbf{j} + x\mathbf{k}) \cdot \mathbf{n}\,ds$, where S is defined by $y = x^2$, $0 \le y \le 4$, $0 \le z \le 1$ and $\mathbf{n}$ is the unit normal to S with positive y-component.

(b) $\int\int_S (yzi_x z\mathbf{j} + xy\mathbf{k}) \cdot \mathbf{n}\,ds$, where S is that part of the surface $z = x^2 + y^2$ cut out by the planes $x = 1, x = -1, y = 1, y = -1$, and $\mathbf{n}$ is the unit lower normal to S.

(c) $\int\int_S (x\mathbf{i}+y\mathbf{j})\cdot\mathbf{n}\,ds$, where S is that part of the surface $z = \sqrt{x^2 + y^2}$ below $z = 1$, and $\mathbf{n}$ is the unit normal to S with negative z-component.

(d) $\int\int_S (x^2y\mathbf{i} + xy\mathbf{j} + z\mathbf{k}) \cdot \mathbf{n}\,ds$, where S is defined by $z = 2 - x^2 - y^2$, $0 \le z$, and $\mathbf{n}$ is the unit normal to S with negative z-component.

(e) $\int\int_S (x\mathbf{i} + y\mathbf{j}) \cdot \mathbf{n}ds$, where S is the surface $x^2 + y^2 + z^2 = 4$, $1 \le z$, and $\mathbf{n}$ is the unit upper normal to S.

(f) $\int\int_S (y\mathbf{i} - x\mathbf{j} + \mathbf{k}) \cdot \mathbf{n}ds$, where S is the smaller surface cut by the plane $y + z = 1$, from the sphere $x^2 + y^2 + z^2 = 1$ and $\mathbf{n}$ is the unit upper normal to S.

18. Evaluate $\int\int_S (y\mathbf{i} - x\mathbf{j} + z\mathbf{k}) \cdot ds$ where

(a) S is that part of $z = 9 - x^2 - y^2$ cut out by $z = 2y$.

(b) S is that part of $z = 2y$ cut out by $z = 9 - x^2 - y^2$, and $\mathbf{n}$ is the unit upper normal to S in each case. [*Hint:* Use polar coordinates with pole at $(0, -1)$].

19. A circular tube $S: x^2 + z^2 = 1$, $0 \le y \le 2$ is a model for a part of an artery. Blood flows through the artery and the force over unit area at any point on the arterial wall is given by

$$\mathbf{F} = e^{-y}\mathbf{n} + \frac{1}{y^2 + 1}\mathbf{j}$$

where $\mathbf{n}$ is the unit outer normal to the arterial wall. Blood diffuses through the wall in such a way that if dS is a small area on S, the amount of diffusion through dS in one second is $\mathbf{F} \cdot \mathbf{n}dS$. Find the total amount of blood leaving the entire wall per second.

20. Evaluate the triple integral over the region given

(a) $\int\int\int_V xy\, dV$, where V is enclosed by $z = \sqrt{1 - x^2 - y^2}$, $z = 0$.

(b) $\int\int\int_V (x+y+z)\, dV$, where V is bounded by $x = 0$, $x = 1$, $z = 0$, $y + z = 2$, $y = z$.

(c) $\int\int\int_V x^2 y\, dV$, where V is the first octant volume bounded by $z = 1$, $z = x^2/4 + y^2/9$.

(d) $\int\int\int_V (x^2 + 2z)\, dV$, where V is bounded by $z = 0$, $y + z = 4$, $y = x^2$.

21. Find the volume bounded by the given surfaces

(a) $z = \sqrt{2 - x^2 - y^2}$, $z = x^2 + y^2$.

(b) $z = x^2 + y^2$, $z = 4 - x^2 - y^2$.

22. Find the volume inside the sphere $x^2 + y^2 + z^2 = 4$ but outside the cylinder $x^2 + y^2 = 1$.

23. Find the volume described.

(a) Bounded by $z = \sqrt{x^2 + y^2}$, $z = \sqrt{1 - x^2 - y^2}$.

(b) Bounded by $x^2 + y^2 + z^2 = 1$, $y = x$, $y = 2x$, $z = 0$ (in first octant).

(c) Bounded by $z = 2\sqrt{x^2 + y^2}$, $x^2 + y^2 = 4$, $z = 0$.

24. (a) What is the integral of the function $x^2 z$ taken over the entire surface of the right circular cylinder of height h which stands on the circle $x^2 + y^2 = a^2$? What is the integral of the given function taken throughout the volume of the cylinder?

(b) Repeat part (a) for the case of the sphere $x^2 + y^2 + z^2 = a^2$

Chapter 6

Integral Theorems

In this chapter we examine the integral theorems of vector analysis that can be extremely useful in the evaluation of various integrals such as those discussed in Chapter 5. There are mainly three integral theorems that will be examined due to their importance in vector analysis applications: Green's theorem, Stokes' theorem and the Divergence theorem (Gauss' theorem). These theorems are very useful in developing numerical or analytical solution techniques arising in different branches of applied science and engineering.

6.1 Green's Theorem

Double integrals over a plane region may be transformed into line integrals over the boundary of the region. Conversely, line integrals over the boundary of a region may be transformed into double integrals over the plane region. This transformation is of practical interest because it may help make the evaluation of an integral easier. It also helps in the theory whenever one wants to switch from one type of integral to another. The transformation can be done by the use of Green's theorem in the plane.

George Green (1793–1841) was a self-educated English mathematician who started out as a baker and at his death was a fellow of Caius College of Cambridge University. His work concerned potential theory in connection with electricity and magnetism, vibrations, waves and elasticity. The work he accomplished remained almost unknown, even in England, until after his death.

Line integrals in the xy-plane are of the form $\int_C f(x, y)\, ds$ and, in the special case that $f(x, y)$ is the tangential component of some vector field $\mathbf{F}(x, y) = P(x, y)\mathbf{i} + Q(x, y)\mathbf{j}$ along C, the integrals take the form

$$\int_C \mathbf{F} \cdot d\mathbf{R} = \int_C P(x, y)\, dx + Q(x, y)\, dy. \qquad (6.1)$$

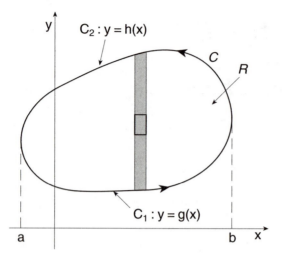

FIGURE 6.1
A simple region R

We now show that when C is a closed curve, the line integral in (6.1) can usually be replaced by a double integral. The precise result is contained in the following theorem, which is sometimes known as Green's lemma.

THEOREM 6.1

If C is a plane region bounded by a finite number of simple closed curves, and if $P(x,y)$, $Q(x,y)$ have continuous first partial derivatives in a domain D containing C and R, where dA is the elementary area of the domain D, then

$$\oint_C P\,dx + Q\,dy = \int\int_R \left(\frac{\partial Q}{\partial x} - \frac{\partial P}{\partial y} \right) dA \qquad (6.2)$$

provided the line integral is taken in the positive direction around C.

Comment: Formula (6.2) can be written in vector form as

$$\int\int_R (\mathbf{curl\ F}) \cdot \mathbf{k}\,dx\,dy = \oint_C \mathbf{F} \cdot d\mathbf{R}$$

This follows from the fact that the third component of the curl of $\mathbf{F}$ is $\partial F_2/\partial x - \partial F_1/\partial y$ (where $\mathbf{F} = F_1\mathbf{i} + F_2\mathbf{j}$).

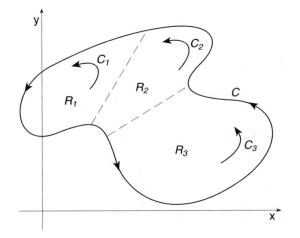

FIGURE 6.2
General region R decomposed in n subregions R_i

PROOF   We first consider a simple region R for which every line parallel to the x- and y-axis that intersects C does so in, at most, two points (Figure 6.1). Then C can be subdivided into an upper and a lower part,

$$C_2 \, : y = h(x) \quad \text{and} \quad C_1 \, : y = g(x).$$

If we consider the second term on the right of (6.2), we have

$$\int\int_R -\frac{\partial P}{\partial y} \, dA = \int_a^b \int_{g(x)}^{h(x)} -\frac{\partial P}{\partial y} \, dy \, dx$$

$$= \int_a^b \{-P\}_{g(x)}^{h(x)} \, dx$$

$$= \int_a^b \{P(x,g(x)) - P(x,h(x))\} \, dx.$$

On the other hand, the first term on the left of (6.2) is

$$\oint_C P \, dx = \int_{C_1} P \, dx + \int_{C_2} P \, dx = \int_{C_1} P \, dx - \int_{-C_2} P \, dx$$

and, if we use x as a parameter along C_1 and $-C_2$, then we have

$$\oint_C P \, dx = \int_a^b P(x, g(x)) \, dx - \int_a^b P(x, h(x)) \, dx$$
$$= \int_a^b \{P(x, g(x)) - P(x, h(x))\} \, dx.$$

Thus, we have shown that

$$\oint_C P \, dx = \int \int_R -\frac{\partial P}{\partial y} \, dA.$$

By subdividing C into two parts of the type $x = g(y)$ and $x = h(y)$ (with $g(y) \le h(y)$), we can also show that

$$\oint_C Q \, dy = \int \int_R \frac{\partial Q}{\partial x} \, dA$$
$$= \int_a^b \int_{g(y)}^{h(y)} \frac{\partial Q}{\partial x} dx \, dy$$
$$= \int_a^b \{Q(h(y), y) - Q(g(y), y)\} \, dy.$$

Addition of these results gives Green's theorem for this C and R.

Consider a more general region R such as that in Figure 6.2; it can be decomposed into n sub-regions R_i, each of which satisfies the condition that lines parallel to the coordinate axes intersect its boundary in at most two points. For each sub-region R_i, Green's theorem gives

$$\oint_{C_i} P \, dx + Q \, dy = \int \int_{R_i} \left(\frac{\partial Q}{\partial x} - \frac{\partial P}{\partial y} \right) \, dA.$$

If these results are added, we get

$$\sum_{i=1}^n \left\{ \oint_{C_i} P \, dx + Q \, dy \right\} = \sum_{i=1}^n \int \int_{R_i} \left(\frac{\partial Q}{\partial x} - \frac{\partial P}{\partial y} \right) \, dA.$$

Now, R is composed of the sub-regions R_i; therefore, the right side of this equation is the double integral over R. Figure 6.2 illustrates that when line integrals over C_i are added, contributions from interior curves cancel in pairs, leaving the line integral around C. This completes the proof. ∎

The proof for even more general regions that cannot be divided into a finite number of sub-regions is the subject of more advanced texts. We continue with further applications of Green's theorem. Later in this chapter

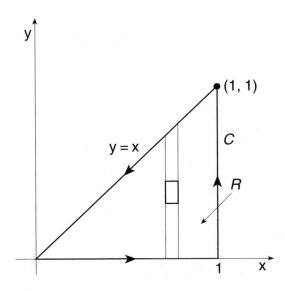

FIGURE 6.3
Line integral for Example 6.1

we will use Green's theorem as the basic tool in deriving two very important theorems (by Gauss and Stokes) for transforming integrals.

Example 6.1
Evaluate the line integral $\oint_C y^2\,dx + x^2\,dy$. ⬜

SOLUTION   By Green's theorem (see Figure 6.3), we have

$$\oint_C y^2\,dx + x^2\,dy = \int\int_R (2x - 2y)\,dA$$

$$= 2\int_0^1 \int_0^x (x - y)\,dy\,dx$$

$$= 2\int_0^1 \left\{ xy - \frac{y^2}{2} \right\}_0^x dx = \int_0^1 x^2\,dx$$

$$= \left\{ \frac{x^3}{3} \right\}_0^1 = \frac{1}{3}.$$

∎

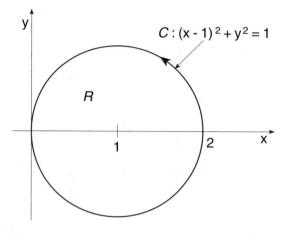

FIGURE 6.4
Circular region C **for Example 6.2**

Example 6.2
Evaluate the integral

$$\oint_C (5x^2 + ye^x + y)\, dx + (e^x + e^y)\, dy,$$

where C is the circle $(x - 1)^2 + y^2 = 1$. []

SOLUTION   By Green's theorem (see Figure 6.4), we again have

$$\oint_C (5x^2 + ye^x + y)\, dx + (e^x + e^y)\, dy = \int\int_R (e^x - e^y - 1)\, dA$$

$$= -\int\int_R dA$$

$$= -(\text{area of } R) = -\pi(1)^2$$

$$= -\pi.$$

Example 6.3
Find the work done by the force $\mathbf{F} = (-16y + \sin x^2)\mathbf{i} + (4e^y + 3x^2)\mathbf{j}$ acting along the simple closed curve C shown in Figure 6.5.   []

SOLUTION   From Chapter 5, the work done by $\mathbf{F}$ is given by

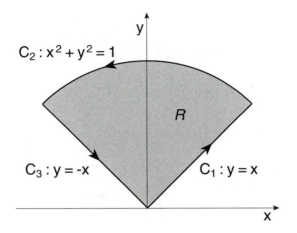

FIGURE 6.5
Simple closed curve for Example 6.3

$$W = \oint_C \mathbf{F} \cdot d\mathbf{R} = \oint_C (-16y + \sin x^2)\, dx + (4e^y + 3x^2)\, dy$$

and so, by Green's theorem

$$W = \int\int_R (6x + 16)\, dA.$$

In view of the region R, the last integral is best handled in polar coordinates. Because R is defined by $0 \le r \le 1$, $\pi/4 \le \theta \le 3\pi/4$,

$$W = \int_{\pi/4}^{3\pi/4} \int_0^1 (6r\cos\theta + 16)r\, dr\, d\theta$$

$$= \int_{\pi/4}^{3\pi/4} (2r^3 \cos\theta + 8r^2)\big|_0^1\, d\theta$$

$$= \int_{\pi/4}^{3\pi/4} (2\cos\theta + 8)\, d\theta = 4\pi.$$

Green's theorem cannot be used to evaluate a line integral around a closed curve that contains a point at which either $P(x,y)$ or $Q(x,y)$ fails to have continuous first partial derivatives. For example, let C be the closed curve consisting of four straight line segments, C_1, C_2, C_3, C_4 shown in Figure 6.6. Green's theorem is *not* applicable to the line integral

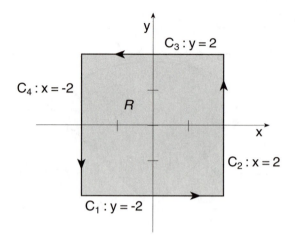

FIGURE 6.6
Straight line segments that make up C

$$\oint \frac{-y}{x^2 + y^2}\, dx + \frac{x}{x^2 + y^2}\, dy$$

because p, Q, $\partial P/\partial y$, and $\partial Q/\partial x$ are not continuous at the origin.

6.2   Region with Holes

Green's theorem can be extended to a region R with "holes," i.e., a region bounded between two or more piecewise smooth simple closed curves. Figure 6.7 (a) shows a region R bounded by a curve C that consists of two simple closed curves, C_1 and C_2, that is, $C = C_1 \cup C_2$. The curve C is positively oriented, because if we traverse C_1 in a counterclockwise direction and C_2 in a clockwise direction, the region R is always to the left. If crosscuts are introduced as shown in Figure 6.7 (b), the region R is divided into two sub-regions, R_1 and R_2. By applying Green's theorem to R_1 and R_2, we obtain

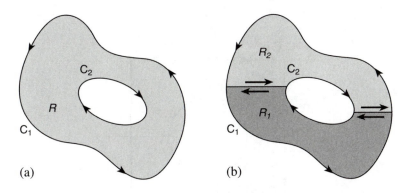

FIGURE 6.7
Regions R bounded by $C = C_1 \cup C_2$

$$\int\int_R \left(\frac{\partial Q}{\partial x} - \frac{\partial P}{\partial y}\right) dA = \int\int_{R_1} \left(\frac{\partial Q}{\partial x} - \frac{\partial P}{\partial y}\right) dA$$

$$+ \int\int_{R_2} \left(\frac{\partial Q}{\partial x} - \frac{\partial P}{\partial y}\right) dA$$

$$= \oint_{C_1} P\,dx + Q\,dy$$

$$+ \oint_{C_2} P\,dx + Q\,dy$$

$$= \oint_C P\,dx + Q\,dy. \qquad (6.3)$$

The last result follows from the fact that the line integrals on the crosscuts (paths with opposite orientations) cancel each other.

Example 6.4
Evaluate

$$\oint_C \frac{-y}{x^2 + y^2}\,dx + \frac{y}{x^2 + y^2}\,dy,$$

⬜

where $C = C_1 \cup C_2$ is the boundary of the shaded region R shown in Figure 6.8.

SOLUTION   Because

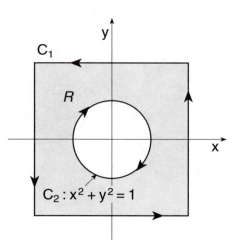

FIGURE 6.8
Region R for Example 6.4

$$P(x,y) = -\frac{y}{x^2+y^2} \quad Q(x,y) = \frac{x}{x^2+y^2}$$

$$\frac{\partial P}{\partial y} = \frac{y^2-x^2}{(x^2+y^2)^2} \quad \frac{\partial Q}{\partial x} = \frac{y^2-x^2}{(x^2+y^2)^2}$$

are continuous on the region R bounded by C, it follows from the preceding discussion that

$$\oint_C \frac{-y}{x^2+y^2}\,dx + \frac{x}{x^2+y^2}\,dy$$

$$= \int\int_R \left\{ \frac{y^2-x^2}{(x^2+y^2)^2} \right.$$

$$\left. - \frac{y^2-x^2}{(x^2+y^2)^2} \right\}\,dA$$

$$= 0.$$

∎

As a consequence of the discussion preceding Example 6.4, we can establish a result for line integrals that enables us, under certain circumstances, to replace a complicated closed path with a path that is simpler.

Suppose, as shown in Figure 6.9, that C_1 and C_2 are two nonintersecting piecewise smooth simple closed paths that have the same counterclockwise

FIGURE 6.9
Two piecewise smooth simple closed paths

orientation. Suppose further that P and Q have continuous first partial derivatives such that

$$\frac{\partial P}{\partial y} = \frac{\partial Q}{\partial x}$$

in the region R bounded between C_1 and C_2. Then from (6.3) we have

$$\oint_{C_1} P\,dx + Q\,dy + \oint_{-C_2} P\,dx + Q\,dy = 0$$

or

$$\oint_{C_1} P\,dx + Q\,dy = \oint_{C_2} P\,dx + Q\,dy. \qquad (6.4)$$

Example 6.5
Evaluate the line integral

$$\oint_C \frac{-y}{x^2 + y^2}\,dx + \frac{x}{x^2 + y^2}\,dy.$$

☐

SOLUTION
One method of evaluating the line integral is to write

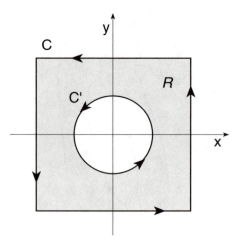

FIGURE 6.10
Region R containing the circle $x^2 + y^2 = 1$

$$\oint_C = \int_{C_1} + \int_{C_2} + \int_{C_3} + \int_{C_4}$$

and then evaluate the four integrals on the line segments C_1, C_2, C_3, and C_4.

Alternatively, if we note that the circle C': $x^2 + y^2 = 1$ lies entirely within C (see Figure 6.10), then from Example 6.4 it is apparent that $P = -y/(x^2 + y^2)$ and $Q = x/(x^2 + y^2)$ have continuous first partial derivatives in the region R bounded between C and C'. In addition,

$$\frac{\partial P}{\partial y} = \frac{y^2 - x^2}{(x^2 + y^2)^2} = \frac{\partial Q}{\partial x}$$

in R. Therefore, it follows from (6.5) that

$$\oint_C \frac{-y}{x^2 + y^2}\, dx + \frac{x}{x^2 + y^2}\, dy = \oint_{C'} \frac{-y}{x^2 + y^2}\, dx + \frac{x}{x^2 + y^2}\, dy.$$

Using the parameterization for C':

$$x = \cos t, \quad y = \sin t, 0 \le t \le 2\pi$$

we obtain

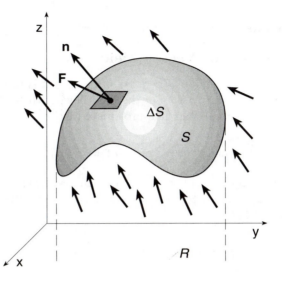

FIGURE 6.11
Surface S with a unit normal to it

$$\oint_C \frac{-y}{x^2 + y^2}\, dx + \frac{x}{x^2 + y^2}\, dy = \int_0^{2\pi} \left(-\sin t(-\sin t) + \cos t(\cos t)\right) dt$$

$$= \int_0^{2\pi} \left(\sin^2 t + \cos^2 t\right) dt$$

$$= \int_0^{2\pi} dt = 2\pi. \tag{6.5}$$

It is interesting to note that the result in (6.6):

$$\oint_C \frac{-y}{x^2 + y^2}\, dx + \frac{x}{x^2 + y^2}\, dy = 2\pi$$

is true for every piecewise smooth simple closed curve C with the origin in its interior. We need only choose C' to be $x^2 + y^2 = a^2$, where a is small enough so that the circle lies entirely within C.

6.3   Integrals over Vector Fields

If $\mathbf{F}(x, y, z) = P(x, y, z)\mathbf{i} + Q(x, y, z)\mathbf{j} + R(x, y, z)\mathbf{k}$ is the velocity field of a fluid element, then the volume of the fluid flowing through an element of surface area ΔS per unit time is approximated by

$$(\text{height})(\text{area of base}) = (\mathbf{F} \cdot \mathbf{n})\Delta S$$

where $\mathbf{n}$ is a unit normal to the surface (see Figure 6.11). The total volume of a fluid passing through S per unit time is called the **flux of F through** S and is given by

$$\int\int_S f(x, y, z)\, dS = \int\int_S (\mathbf{F} \cdot \mathbf{n})\, dS. \qquad (6.6)$$

In the case of a closed surface S, if $\mathbf{n}$ is the outer(inner) normal, then (6.7) gives the volume of fluid flowing out(in) through S per unit time. Note in the integral given by (6.7) that $f(x, y, z)$ is the component of $\mathbf{F}$ in the direction of $\mathbf{n}$.

Example 6.6

Let $\mathbf{F}(x, y, z) = z\mathbf{j} + z\mathbf{k}$ represent the flow of a liquid. Find the flux of $\mathbf{F}$ through the surface S given by that portion of the plane $z = 6 - 3x - 2y$ in the first octant oriented upward.   ⬜

SOLUTION   The vector field and the surface are illustrated in Figure 6.12. By defining the plane $g(x, y, z) = 3x + 2y + z - 6 = 0$, we see that a unit normal with a positive $\mathbf{k}$ component is

$$\mathbf{n} = \frac{\nabla g}{|\nabla g|} = \frac{3}{\sqrt{14}}\mathbf{i} + \frac{2}{\sqrt{14}}\mathbf{j} + \frac{1}{\sqrt{14}}\mathbf{k}.$$

Hence,

$$\text{flux} = \int\int_S f(x, y, z)\, dS = \int\int_S (\mathbf{F} \cdot \mathbf{n})\, dS = \frac{1}{\sqrt{14}} \int\int_S 3z\, dS$$

with R the projection of the surface onto the xy-plane, we find from (6.7) that

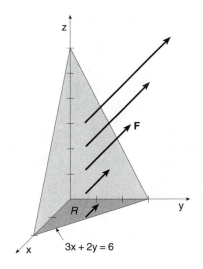

FIGURE 6.12
Surface S with flux F for Example 6.6

$$\int\int_S (\mathbf{F} \cdot \mathbf{n})\, dS = \frac{1}{\sqrt{14}} \int\int_R 3(6 - 3x - 2y)(\sqrt{14}\, dA)$$

$$= 3 \int_0^2 \int_0^{3-3x/2} (6 - 3x - 2y) dy\, dx = 18.$$

∎

Depending on the nature of the vector field, the integral (6.7) can represent other kinds of flux. For example (6.7) could also give electric flux, magnetic flux, flux of heat, and so on.

REMARK 6.1   If the surface S is piecewise defined, we express a surface integral over S as the sum of the surface integrals over the various pieces of the surface. For example, suppose S is the orientable piecewise smooth closed surface bounded by the paraboloid $z = x^2 + y^2$ (S_1) and the plane $z = 1$ (S_2). Then the flux of a vector field $\mathbf{F}$ out of the surface S is

$$\int\int_S \mathbf{F} \cdot \mathbf{n}\, dS = \int\int_{S_1} \mathbf{F} \cdot \mathbf{n}\, dS + \int\int_{S_2} \mathbf{F} \cdot \mathbf{n}\, dS$$

where we take S_1 oriented upward and S_2 oriented downward (see Figure 6.13)   ∎

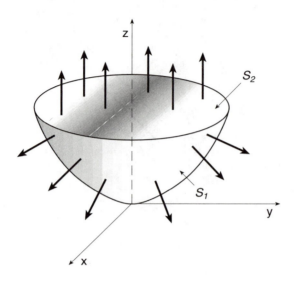

FIGURE 6.13
Orientable, piecewise smooth, closed surface S

Example 6.7
Evaluate

$$\int \int_S \mathbf{F} \cdot \mathbf{n}\, dS$$

where $\mathbf{F} = x\mathbf{i} + y\mathbf{j} + z\mathbf{k}$ and $\mathbf{n}$ is the unit outward-pointing normal to the surface S enclosing the volume bounded by $x^2 + y^2 = 4$, $z = 0$, $z = 2$. $\square$

SOLUTION   We divide S into four parts (Figure 6.14):

$$S_1 : z = 0, \quad x^2 + y^2 \le 4;$$
$$S_2 : z = 2, \quad x^2 + y^2 \le 4;$$
$$S_3 : y = \sqrt{4 - x^2}, \quad 0 \le z \le 2;$$
$$S_4 : y = -\sqrt{4 - x^2}, \quad 0 \le z \le 2.$$

On S_1, $\mathbf{n} = -\mathbf{k}$; on S_2, $\mathbf{n} = \mathbf{k}$; and on S_3 and S_4

$$\mathbf{n} = \frac{\nabla(x^2 + y^2 - 4)}{|\nabla(x^2 + y^2 - 4)|} = \frac{(2x, 2y, 0)}{\sqrt{4x^2 + 4y^2}} = \frac{(x, y, 0)}{2}.$$

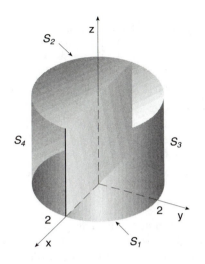

FIGURE 6.14
Surface S divided into four regions for Example 6.7

We know that S_1 and S_2 project onto S_{xy} : $x^2 + y^2 \le 4$ in the xy-plane, and S_3 and S_4 project onto the rectangle S_{xz} : $-2 \le x \le 2$, $0 \le z \le 2$ in the xz-plane. Consequently,

$$
\iint_S \mathbf{F} \cdot \mathbf{n}\, dS = \iint_{S_1} -z\, dS + \iint_{S_2} d\, dS + \iint_{S_3} \left(\frac{x^2 + y^2}{2}\right) dS
$$

$$
+ \iint_{S_4} \left(\frac{x^2 + y^2}{2}\right) dS
$$

$$
= 0 + \iint_{S_{xy}} 2\, dA
$$

$$
+ \frac{1}{2} \iint_{S_{xz}} (x^2 + 4 - x^2)\sqrt{1 + \left(\frac{-x}{\sqrt{4 - x^2}}\right)^2}\, dA
$$

$$
+ \frac{1}{2} \iint_{S_{xz}} (x^2 + 4 - x^2)\sqrt{1 + \left(\frac{x}{\sqrt{4 - x^2}}\right)^2}\, dA
$$

$$
= 2 \iint_{S_{xy}} dA + 4 \iint_{S_{xz}} \frac{2}{\sqrt{4 - x^2}}\, dA
$$

$$
= 2(\text{area of } S_{xy}) + 8 \int_{-2}^{2} \int_0^2 \frac{1}{\sqrt{4 - x^2}}\, dz\, dx
$$

$$
= 2(4\pi) + 16 \int_{-2}^{2} \frac{1}{\sqrt{4 - x^2}}\, dx.
$$

If we set $x = 2 \sin \theta$, then $dx = 2 \cos \theta d\theta$ and

$$\int \int_S \mathbf{F} \cdot \mathbf{n} \, dS = 8\pi + 16 \int_{-\pi/2}^{\pi/2} \frac{1}{2 \cos \theta} 2 \cos \theta \, d\theta$$
$$= 8\pi + 16 \left\{\theta\right\}_{-\pi/2}^{\pi/2} = 24\pi.$$

∎

Example 6.8

If a spherical object is submerged in a fluid of density ρ, it experiences a buoyant force due to fluid pressure. Show that the magnitude of this force is exactly the weight of fluid displaced by the object.   ⬚

SOLUTION   Suppose the object is represented by the sphere $x^2 + y^2 + z^2 = R^2$ and the surface of the fluid by the plane $z = h$ ($h > R$) (see Figure 6.15). If dS is a small area on the surface, then the force due to fluid pressure on dS has magnitude $P \, dS$, where P is pressure, and this force acts normal to the surface of the sphere. If $\mathbf{n}$ is the unit inward-pointing normal to the sphere, then the force on dS is

$$(P \, dS)\mathbf{n}.$$

Clearly, the resultant force on the sphere will be in the z-direction, the y-components cancelling because of the symmetry of the sphere. The z-component of the force on dS is $(P \, dS)\mathbf{n} \cdot \mathbf{k}$; thus the magnitude of the resultant force on the sphere is

$$\int \int_S P \mathbf{k} \cdot \mathbf{n} \, dS.$$

A normal to the surface is $\nabla(x^2 + y^2 + z^2 - R^2) = (2x, 2y, 2z)$, and therefore the unit inward-pointing normal is

$$\mathbf{n} = -\frac{(x, y, z)}{\sqrt{x^2 + y^2 + z^2}} = -\frac{1}{R}(x, y, z).$$

Using spherical coordinates (see Chapter 5) for an element on the sphere, and $P = 9.81\rho(h - z)$,

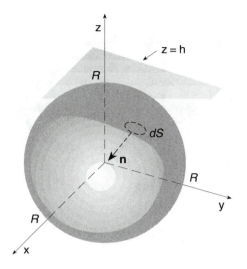

FIGURE 6.15
Object represented by the sphere and surface of fluid by a plane

$$\int\int_S P\mathbf{k}\cdot\mathbf{n}\,dS = \int\int_S 9.81\rho(h-z)\left(-\frac{z}{R}\right)dS$$

$$= -\frac{9.81\rho}{R}\int_0^{2\pi}\int_0^{\pi}(h-R\cos\phi)(R\cos\phi)R^2\sin\phi\,d\phi\,d\theta$$

$$= -9.81\rho R^2\int_0^{2\pi}\left\{-\frac{h}{2}\cos^2\phi+\frac{R}{3}\cos^3\phi\right\}_0^{\pi}d\theta$$

$$= \frac{2(9.81)\rho R^3}{3}\{\theta\}_0^{2\pi}$$

$$= \frac{4}{3}\pi R^3(9.81\rho).$$

6.4   Stokes' Theorem

Green's theorem of the preceding section has two vector forms. In this and the next section we shall generalize these forms to three dimensions. If $\mathbf{F}(x,y) = P(x,y)\mathbf{i} + Q(x,y)\mathbf{j}$ is a two-dimensional vector field, then

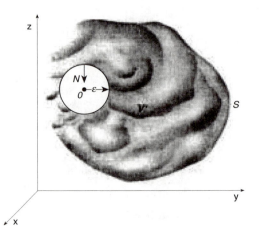

FIGURE 6.16
Singular point excluded from three-dimensional region

$$\textbf{curl F} = \nabla \times \textbf{F} = \begin{vmatrix} \textbf{i} & \textbf{j} & \textbf{k} \\ \frac{\partial}{\partial x} & \frac{\partial}{\partial y} & \frac{\partial}{\partial z} \\ P & Q & 0 \end{vmatrix}$$

$$= \left(\frac{\partial Q}{\partial x} - \frac{\partial P}{\partial y} \right) \textbf{k}.$$

From Chapter 5, we can write Green's theorem in vector notation as

$$\oint_C \textbf{F} \cdot d\textbf{R} = \oint_C \textbf{F} \cdot \textbf{T} \, ds = \int \int_R (\nabla \times \textbf{F}) \cdot \textbf{k} \, dA \qquad (6.7)$$

that is, the line integral of the tangential component of $\textbf{F}$ is the double
integral of the normal component of the curl of $\textbf{F}$ ($\nabla \times \textbf{F}$).

6.5   Green's Theorem in 3-D

The vector form of Green's theorem given in (6.8) relates a line integral
around a piecewise smooth simple closed curve C, forming the boundary of a
plane region R to a double integral over R. Green's theorem in 3-space relates
a line integral around a piecewise smooth simple closed curve C forming the

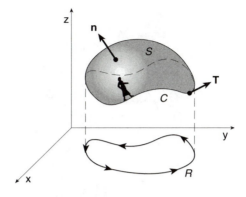

FIGURE 6.17
Region R showing orientation of surface

boundary of a surface S with a surface integral over S. Suppose $z = f(x, y)$ is a continuous function whose graph is a piecewise smooth orientable surface over a region R on the xy-plane.

Let C form the boundary of S and let the projection of C onto the xy-plane form the boundary of R. The positive direction on C is induced by the orientation of the surface S; the positive direction on C corresponds to the direction a person would have to walk on C to have his or her head pointing in the direction of the orientation of the surface, while keeping the surface to the left (see Figure 6.17). More precisely, the positive orientation of C is in accordance with the right-hand rule: If the thumb of the right hand points in the direction of the orientation of the surface, then roughly the fingers of the right hand wrap around the surface in the positive direction. Finally, let $\mathbf{T}$ be a unit tangent vector to C that points in the positive direction.

The three-dimensional form of Green's theorem is known as Stokes' theorem, after the Irish mathematician and physicist George G. Stokes (1819–1903). Like George Green, George Stokes was a don at Cambridge University. In 1854, Stokes posed this theorem as a problem on a prize examination for Cambridge students.

THEOREM 6.2
Let C be a closed, piecewise-smooth curve that does not intersect itself and let S be a piecewise-smooth, orientable surface with C as boundary (Figure 6.18). Let $\mathbf{F}(x, y, z) = P(x, y, z)\mathbf{i} + Q(x, y, z)\mathbf{j} + R(x, y, z)\mathbf{k}$ be a vector field whose components P, Q, and R have continuous first partial derivatives in a domain that contains S and C. Then

$$\oint_C \mathbf{F} \cdot d\mathbf{R} = \oint_C \mathbf{F} \cdot \mathbf{T}\, ds = \int\int_S (\nabla \times \mathbf{F}) \cdot \mathbf{n}\, dS \qquad (6.8)$$

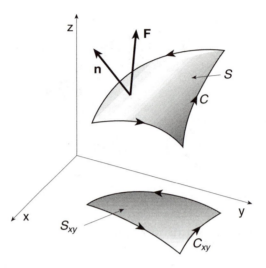

FIGURE 6.18
Piecewise smooth curve C that does not intersect itself

or

$$\oint_C P\,dx + Q\,dy + R\,dz = \int\int_S \left\{ \left(\frac{\partial R}{\partial y} - \frac{\partial Q}{\partial z} \right) \mathbf{i} \right. $$
$$+ \left(\frac{\partial P}{\partial z} - \frac{\partial R}{\partial x} \right) \mathbf{j} $$
$$\left. + \left(\frac{\partial Q}{\partial x} - \frac{\partial P}{\partial y} \right) \mathbf{k} \right\} \cdot \mathbf{n}\,dS \qquad (6.9)$$

where $\mathbf{n}$ is the unit normal to S chosen in the following way: If, when moving along C the surface S is on the left, then $\mathbf{n}$ must be chosen as the unit normal on that side of S. On the other hand if, when moving along C, the surface is on the right, then $\mathbf{n}$ must be chosen on the opposite side of S.

PROOF   We first consider a surface S that project in a one-to-one fashion onto each of the three coordinate planes. Because S projects one-to-one onto some region S_{xy} in the xy-plane, we can take the equation for S in the form $z = f(x, y)$. If the direction along C is as indicated in Figure 6.18, and C_{xy} is the projection of C on the xy-plane, then

$$\oint_C P\,dx = \oint_{C_{xy}} P\left[x, y, f(x, y)\right]\,dx.$$

If we use Green's theorem on this line integral around C_{xy}, we have

$$\oint_C P \, dx = \int \int_{S_{xy}} - \left\{ \frac{\partial P}{\partial y} + \frac{\partial P}{\partial z} \frac{\partial z}{\partial y} \right\} dA.$$

On the other hand, because a unit normal to S is

$$\mathbf{n} = \frac{\left(-\frac{\partial z}{\partial x}, -\frac{\partial z}{\partial y}, 1 \right)}{\sqrt{1 + \left(\frac{\partial z}{\partial x} \right)^2 + \left(\frac{\partial z}{\partial y} \right)^2}}.$$

it also follows that

$$\int \int_S \left(\frac{\partial P}{\partial z} \mathbf{j} - \frac{\partial P}{\partial y} \mathbf{k} \right) \cdot \mathbf{n} \, dS = \int \int_S \frac{\left(-\frac{\partial P}{\partial z} \frac{\partial z}{\partial y} - \frac{\partial P}{\partial y} \right)}{\sqrt{1 + \left(\frac{\partial z}{\partial x} \right)^2 + \left(\frac{\partial z}{\partial y} \right)^2}} \, dS$$

$$= \int \int_{S_{xy}} \frac{\left(-\frac{\partial P}{\partial z} \frac{\partial z}{\partial y} - \frac{\partial P}{\partial y} \right)}{\sqrt{1 + \left(\frac{\partial z}{\partial x} \right)^2 + \left(\frac{\partial z}{\partial y} \right)^2}}$$

$$\times \sqrt{1 + \left(\frac{\partial z}{\partial x} \right)^2 + \left(\frac{\partial z}{\partial y} \right)^2} \, dA$$

$$= \int \int_{S_{xy}} - \left\{ \frac{\partial P}{\partial y} + \frac{\partial P}{\partial z} \frac{\partial z}{\partial y} \right\} dA.$$

We have shown, then, that

$$\oint_C P \, dx = \int \int_S \left(\frac{\partial P}{\partial z} \mathbf{j} - \frac{\partial P}{\partial y} \mathbf{k} \right) \cdot \mathbf{n} \, dS.$$

By projecting C and S onto the xz- and yz-planes, we can similarly show that

$$\oint_C R \, dz = \int \int_S \left(\frac{\partial P}{\partial y} \mathbf{i} - \frac{\partial R}{\partial x} \mathbf{j} \right) \cdot \mathbf{n} \, dS$$

and

$$\oint_C Q \, dy = \int \int_S \left(\frac{\partial Q}{\partial x} \mathbf{k} - \frac{\partial Q}{\partial z} \mathbf{i} \right) \cdot \mathbf{n} \, dS.$$

Addition of these results gives Stokes' theorem for $\mathbf{F}$ and S and completes the proof. ∎

   The proof can be extended to more general curves and surfaces that do not project in a one-to-one fashion onto all three coordinate planes. Most

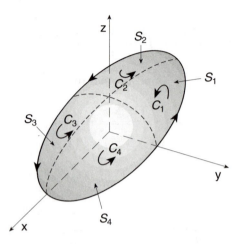

FIGURE 6.19
Line integrals over C_i cancelling out in pairs

surfaces S with bounding curves C can be subdivided into n subsurfaces S_i with bounding curves C_i that do satisfy this condition (Figure 6.19). For each such sub-surface, Stokes' theorem applies

$$\oint_{C_i} \mathbf{F} \cdot d\mathbf{R} = \int\int_{S_i} (\nabla \times \mathbf{F}) \cdot \mathbf{n}\, dS, \quad i = 1, \ldots, n.$$

If these n equations are now added together, we have

$$\sum_{i=1}^n \oint_{C_i} \mathbf{F} \cdot d\mathbf{R} = \sum_{i=1}^n \int\int_{S_i} (\nabla \times \mathbf{F}) \cdot \mathbf{n}\, dS.$$

Because the S_i constitute S, the right side of this equation is the surface integral of $(\nabla \times \mathbf{F}) \cdot \mathbf{n}$ over S. Figure 6.19 illustrates that when line integrals over C_i are added, contributions from interior curves cancel in pairs, and the remaining line integrals give the line integral $\mathbf{F}d\mathbf{r}$ along C. For general surfaces that cannot be divided into a finite number of sub-surfaces of this type, the reader should consult a more advanced text.

Green's theorem is a special case of Stokes' theorem. If $\mathbf{F}(x, y) = P(x, y)\mathbf{i} + Q(x, y)\mathbf{j}$, and C is a closed curve in the xy-plane, then by Stokes' theorem

$$\oint_C P\,dx + Q\,dy = \int\int_S \left(\frac{\partial Q}{\partial x} - \frac{\partial P}{\partial y} \right) \mathbf{k} \cdot \mathbf{n}\, dS$$

where S is any surface for which C is the boundary. If we choose S as that part of the xy-plane bounded by C, then $\mathbf{n} = \mathbf{k}$ and

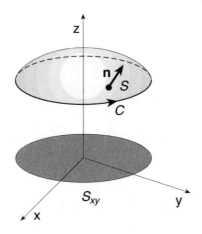

FIGURE 6.20
Surface S with upper normal n for Example 6.9

$$\oint_C P\,dx + Q\,dy = \int\int_S \left(\frac{\partial Q}{\partial x} - \frac{\partial P}{\partial y}\right)\,dA.$$

Suppose that the curl of a vector field **F** vanishes in a simply connected domain D. If C is any piecewise-smooth, closed curve in D, then there exists a piecewise-smooth surface S in D with C as boundary. By Stokes' theorem

$$\oint_C \mathbf{F} \cdot d\mathbf{R} = \int\int (\nabla \times \mathbf{F}) \cdot \mathbf{n}\,dS = 0.$$

This implies that the line integral is independent of path in D, and that there exists a function $f(x, y, z)$ such that $\nabla f = \mathbf{F}$.

Example 6.9
Verify Stokes' theorem if $\mathbf{F} = x^2\mathbf{i} + x\mathbf{j} + xyz\mathbf{k}$, and S is that part of the sphere $x^2 + y^2 + z^2 = 4$ above the plane $z = 1$. ▯

SOLUTION   If we choose **n** as the upper normal to s, then C, the boundary of S, must be traversed in the direction shown in Figure 6.20. (If **n** is chosen as the lower normal, then C must be traversed in the opposite direction.) Because parametric equations for C are

$$x\sqrt{3}\cos t, \quad y = \sqrt{3}\sin t, \quad z = 1, \quad 0 \le t \le 2\pi$$

we have

$$\oint_C \mathbf{F} \cdot d\mathbf{r} = \oint_C x^2 \, dx + x \, dy + xyz \, dz$$

$$= \int_0^{2\pi} \left\{ 3 \cos^2 t (-\sqrt{3} \sin t \, dt) + \sqrt{3} \cos t (\sqrt{3} \cos t \, dt) \right\}$$

$$= \int_0^{2\pi} \left\{ -3\sqrt{3} \cos^2 t \sin t + \frac{3}{2}(1 + \cos 2t) \right\} dt$$

$$= \left\{ \sqrt{3} \cos^2 t + \frac{3t}{2} + \frac{3 \sin 2t}{4} \right\}_0^{2\pi}$$

$$= 3\pi.$$

Alternatively,

$$\int \int_S (\nabla \times \mathbf{F}) \cdot \mathbf{n} \, dS = \int \int_S (xz\mathbf{i} - yz\mathbf{j} + \mathbf{k}) \cdot \frac{(2x, 2y, 2z)}{\sqrt{4x^2 + 4y^2 + 4z^2}} \, dS$$

$$= \int \int_S \frac{x^2 z - y^2 z + z}{\sqrt{x^2 + y^2 + z^2}} \, dS$$

$$= \int \int_{S_{xy}} \frac{z(x^2 - y^2 + 1)}{2}$$

$$\times \sqrt{1 + \left(\frac{\partial z}{\partial x} \right)^2 + \left(\frac{\partial z}{\partial y} \right)^2} \, dA$$

$$= \frac{1}{2} \int \int_{S_{xy}} z(x^2 - y^2 + 1)$$

$$\times \sqrt{1 + (-x/z)^2 + (-y/z)^2} \, dA$$

$$= \frac{1}{2} \int \int_{S_{xy}} z(x^2 - y^2 + 1) \frac{\sqrt{x^2 + y^2 + z^2}}{z} \, dA$$

$$= \int \int_{S_{xy}} (x^2 - y^2 + 1) \, dA.$$

If we use polar coordinates to evaluate this double integral over S_{xy}:$x^2 + y^2 \le 3$, we have

$$\int\int_S (\nabla \times \mathbf{F}) \cdot \mathbf{n}\, dS = \int_{-\pi}^{\pi} \int_0^{\sqrt{3}} (r^2 \cos^2 \theta - r^2 \sin^2 \theta + 1) r\, dr\, d\theta$$

$$= \int_{-\pi}^{\pi} \left\{ \frac{r^4}{4}(\cos^2 \theta - \sin^2 \theta) + \frac{r^2}{2} \right\}_0^{\sqrt{3}} d\theta$$

$$= \int_{-\pi}^{\pi} \left\{ \frac{9}{4}\cos 2\theta + \frac{3}{2} \right\} d\theta$$

$$= \left\{ \frac{9}{8}\sin 2\theta + \frac{3\theta}{2} \right\}_{-\pi}^{\pi} = 3\pi.$$

∎

Example 6.10

Let S be the part of the cylinder $z = 1 - x^2$ for $0 \le x \le 1$, $-2 \le y \le 2$. Verify Stokes' theorem if $\mathbf{F} = xy\mathbf{i} + yz\mathbf{j} + xz\mathbf{k}$. ⬚

SOLUTION   The surface S, the curve C (which is composed of the union of C_1, C_2, C_3, and C_4), and the region R are shown in Figure 6.21.

The Surface Integral:

From $\mathbf{F} = xy\mathbf{i} + yz\mathbf{j} + xz\mathbf{k}$, we find

$$\nabla \times \mathbf{F} = \begin{vmatrix} \mathbf{i} & \mathbf{j} & \mathbf{k} \\ \frac{\partial}{\partial x} & \frac{\partial}{\partial y} & \frac{\partial}{\partial z} \\ xy & yz & xz \end{vmatrix} = -y\mathbf{i} - z\mathbf{j} - x\mathbf{k}.$$

Now, if $g(x, y, z) = z + x^2 - 1 = 0$ defines the cylinder, then the upper normal is

$$\mathbf{n} = \frac{\nabla g}{|\nabla g|} = \frac{2x\mathbf{i} + \mathbf{k}}{\sqrt{4x^2 + 1}}.$$

Therefore,

$$\int\int_S (\nabla \times \mathbf{F}) \cdot \mathbf{n}\, dS = \int\int_S \frac{(-2xy - x)}{\sqrt{4x^2 + 1}}\, dS.$$

To evaluate the latter surface integral, we have the following (see Chapter 5):

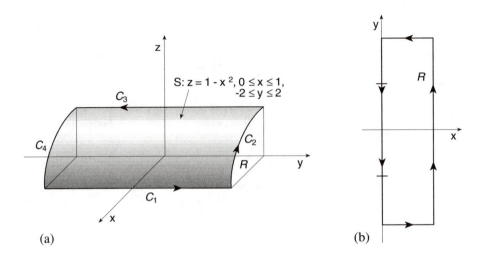

FIGURE 6.21
Surface S and curve C

$$\int\int_S \frac{(-2xy - x)}{\sqrt{4x^2 + 1}} \, dS = \int\int_R (-2xy - x) \, dA$$

$$= \int_0^1 \int_{-2}^2 (-2xy - x) \, dy \, dx$$

$$= \int_0^1 \left(-xy^2 - xy\right)\Big|_{-2}^2 \, dx$$

$$= \int_0^1 (-4x) \, dx = -2. \tag{6.10}$$

The Line Integral:

We write $\oint_C = \int_{C_1} + \int_{C_2} + \int_{C_3} + \int_{C_4}$. On C_1: $x = 0$, $z = 0$, $dx = 0$, $dz = 0$, so

$$\int_{C_1} y(0) + y(0)dy + 0 = 0.$$

On C_2 : $y = 2$, $z = 1 - x^2$, $dy = 0$, $dz = -2x\,dx$, so

$$\int_{C_2} 2x\,dx + 2(1-x^2)(0) + x(1-x^2)(-2x\,dx)$$

$$= \int_1^0 (2x - 2x^2 + 2x^4)\,dx$$

$$= -\frac{11}{15}.$$

On $C_3 : x = 0,\ z = 1,\ dx = 0,\ dz = 0$, so

$$\int_{C_3} 0 + y\,dy + 0 = \int_2^{-2} y\,dy = 0.$$

On $C_4 : y = -2,\ z = 1 - x^2,\ dy = 0,\ dz = -2x\,dx$, so

$$\int_{C_4} -2x\,dx - 2(1-x^2)(0) + x(1-x^2)(-2x\,dx)$$

$$= \int_0^1 (-2x - 2x^2 + 2x^4)\,dx$$

$$= -\frac{19}{15}.$$

Hence,

$$\oint_C xy\,dx + yz\,dy + xz\,dz = 0 - \frac{11}{15} + 0 - \frac{19}{15} = -2$$

which, of course, agrees with (6.11). ∎

Example 6.11
Evaluate the integral

$$\oint_C 2xy^3\,dx + 3x^2y^2\,dy + (2z + z)\,dz$$

where C consists of line segments joining $A(2,0,0)$ to $B(0,1,0)$ to $D(0,0,1)$ to A. ⬚

SOLUTION   By Stokes' theorem,

$$\oint_C 2xy^3\,dx + 3x^2y^2\,dy + (2z + x)\,dz = \int\int_S \nabla \times (2xy^3, 3x^2y^2, 2z + x) \cdot \mathbf{n}\,dS$$

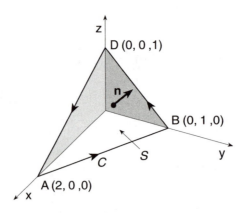

FIGURE 6.22
Flat triangle S bounded by C

where S is any surface with C as boundary. If we choose S as the flat triangle
bounded by C (Figure 6.22), then a normal vector to S is

$$\mathbf{BD} \times \mathbf{BA} = \begin{vmatrix} \mathbf{i} & \mathbf{j} & \mathbf{k} \\ 0 & -1 & 1 \\ 2 & -1 & 0 \end{vmatrix} = (1, 2, 2)$$

and therefore

$$\mathbf{n} = \frac{(1, 2, 2)}{3}.$$

Because

$$\nabla \times (2xy^3, 3x^2y^2, 2z + x) = \begin{vmatrix} \mathbf{i} & \mathbf{j} & \mathbf{k} \\ \frac{\partial}{\partial x} & \frac{\partial}{\partial y} & \frac{\partial}{\partial z} \\ 2xy^3 & 3x^2y^2 & 2z + x \end{vmatrix} = -\mathbf{j}$$

it follows that

$$\nabla \times (2xy^3, 3x^2y^2, 2z + x) \cdot \mathbf{n} = -\mathbf{j} \cdot \frac{(1, 2, 2)}{3} = -\frac{2}{3}$$

and

$$\oint_C 2xy^3 \, dx + 3x^2y^2 \, dy + (2z + x) \, dz = \int \int_S -\frac{2}{3} dS$$

$$= -\frac{2}{3}(\text{area of } S).$$

The area of triangle S is given by

$$\frac{1}{2}|\mathbf{BD} \times \mathbf{BA}| = \frac{1}{2}|(1,2,2)| = \frac{3}{2}.$$

Finally, then

$$\oint_C 2xy^3\, dx + 3x^2y^2\, dy + (2z+x)\, dz = -\frac{2}{3}\left(\frac{3}{2}\right) = -1.$$

∎

6.6 The Divergence Theorem

Another Vector Form of Green's Theorem

Let $\mathbf{F} = P\mathbf{i} + Q\mathbf{j}$ and let $\mathbf{T} = (dx/ds)\mathbf{i} + (dy/ds)\mathbf{j}$ be a unit tangent to a simple closed plane curve C. In the last section we saw that $\oint_C (\mathbf{F} \cdot \mathbf{T})dS$ can be evaluated by a double integral involving the curl of $\mathbf{F}$. Similarly, if $\mathbf{n} = (dy/ds)\mathbf{i} - (dx/ds)\mathbf{j}$ is a unit normal to C (check $\mathbf{T} \cdot \mathbf{n}$), then $\oint_C (\mathbf{F} \cdot \mathbf{n})dS$ can be expressed in terms of a double integral of the divergence of $\mathbf{F}$. From Green's theorem,

$$\oint_C (\mathbf{F} \cdot \mathbf{n})\, dS = \oint_C P\, dy - Q\, dx = \int\int_R \left\{ \frac{\partial P}{\partial x} - \left(-\frac{\partial Q}{\partial y} \right) \right\} dA$$
$$= \int\int_R \left\{ \frac{\partial P}{\partial x} + \frac{\partial Q}{\partial y} \right\} dA$$

that is,

$$\oint_C (\mathbf{F} \cdot \mathbf{n})\, dS = \int\int_R \nabla \cdot \mathbf{F}\, dA. \qquad (6.11)$$

The result in (6.12) is a special case of the **divergence** or Gauss' theorem.

The divergence theorem relates certain surface integrals that enclose volumes to triple integrals over the enclosed volume. More precisely, we have the following.

THEOREM 6.3

Let S be a piecewise-smooth surface enclosing a region V (Figure 6.23 (a)). Let $\mathbf{F}(x,y,z) = L(x,y,z)\mathbf{i} + M(x.y.z)\mathbf{j} + N(x,y,z)\mathbf{k}$ be a vector field whose

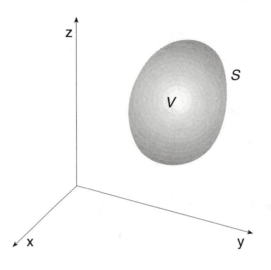

FIGURE 6.23
Piecewise smooth surface V and surface S

components L, M, and N have continuous first partial derivatives in a do-
main containing S and V. If $\mathbf{n}$ is the unit outer normal to S, then

$$\int\int_S \mathbf{F}\cdot\mathbf{n}\,dS = \int\int\int_V \nabla\cdot\mathbf{F}\,dV \qquad (6.12)$$

or

$$\int\int_S (L\mathbf{i}+M\mathbf{j}+N\mathbf{k})\cdot\mathbf{n}\,dS = \int\int\int_V \left(\frac{\partial L}{\partial x}+\frac{\partial M}{\partial y}+\frac{\partial N}{\partial z}\right)dV. \quad (6.13)$$

PROOF   We first consider a surface S for which any line parallel to
any coordinate axis intersects S in at most two points (Figure 6.23 (b)).
We can then divide S into an upper and a lower portion, $S_2 :\ z = f_2(x,y)$
and $S_1 :\ z = f_1(x,y)$, both of which have the same projection S_{xy} in the
xy-plane. We consider the third term in the surface integral on the left side
of (6.14):

$$\int\int_R N\mathbf{k}\cdot\mathbf{n}\,dS = \int\int_{S_1} N\mathbf{k}\cdot\mathbf{n}\,dS + \int\int_{S_2} N\mathbf{k}\cdot\mathbf{n}\,dS. \qquad (6.14)$$

On S_1,

$$\mathbf{n} = \frac{\left(\frac{\partial f_1}{\partial x},\frac{\partial f_1}{\partial y},-1\right)}{\sqrt{1+\left(\frac{\partial f_1}{\partial x}\right)^2+\left(\frac{\partial f_1}{\partial y}\right)^2}}.$$

On S_2,

$$\mathbf{n} = \frac{(\frac{\partial f_2}{\partial x}, \frac{\partial f_2}{\partial y}, -1)}{\sqrt{1 + \left(\frac{\partial f_2}{\partial x}\right)^2 + \left(\frac{\partial f_2}{\partial y}\right)^2}},$$

consequently,

$$\int\int_S N\mathbf{k} \cdot \mathbf{n}\, dS = \int\int_{S_1} \frac{-N}{\sqrt{1 + \left(\frac{\partial f_1}{\partial x}\right)^2 + \left(\frac{\partial f_1}{\partial y}\right)^2}}\, dS$$

$$+ \int\int_{S_2} \frac{N}{\sqrt{1 + \left(\frac{\partial f_2}{\partial x}\right)^2 + \left(\frac{\partial f_2}{\partial y}\right)^2}}\, dS$$

$$= \int\int_{S_{xy}} \frac{-N\{x, y, f_1(x, y)\}}{\sqrt{1 + \left(\frac{\partial f_1}{\partial x}\right)^2 + \left(\frac{\partial f_1}{\partial y}\right)^2}}$$

$$\times \sqrt{1 + \left(\frac{\partial f_1}{\partial x}\right)^2 + \left(\frac{\partial f_1}{\partial y}\right)^2}\, dA$$

$$+ \int\int_{S_{xy}} \frac{N\{x, y, f_2(x, y)\}}{\sqrt{1 + \left(\frac{\partial f_2}{\partial x}\right)^2 + \left(\frac{\partial f_2}{\partial y}\right)^2}}$$

$$\times \sqrt{1 + \left(\frac{\partial f_2}{\partial x}\right)^2 + \left(\frac{\partial f_2}{\partial y}\right)^2}\, dA$$

$$= \int\int_{S_{xy}} \{N[x, y, f_2(x, y)] - N[x, y, f_1(x, y)]\}\, dA.$$

On the other hand,

$$\int\int\int_V \frac{\partial N}{\partial z}\, dV = \int\int_{S_{xy}} \left\{\int_{f_1(x,y)}^{f_2(x,y)} \frac{\partial N}{\partial z}\, dz\right\} dA$$

$$= \int\int_{S_{xy}} \{N\}_{f_1(x,y)}^{f_2(x,y)}\, dA$$

$$= \int\int_{S_{xy}} \{N[x, y, f_2(x, y)] - N[x, y, f_1(x, y)]\}\, dA.$$

We have just shown that

$$\int\int_S N\mathbf{k} \cdot \mathbf{n}\, dS = \int\int\int_V \frac{\partial N}{\partial z}\, dV.$$

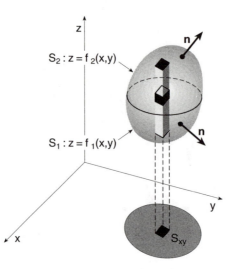

FIGURE 6.24
Addition of surface integrals S_i

Projections of S onto the xz- and yz-planes lead in a similar way to

$$\int\int_S M\mathbf{j}\cdot\mathbf{n}\,dS = \int\int\int_V \frac{\partial M}{\partial y}\,dV$$

$$\text{and } \int\int_S L\mathbf{i}\cdot\mathbf{n}\,dS = \int\int\int_V \frac{\partial L}{\partial x}\,dV.$$

By adding these three results, we obtain the divergence theorem for $\mathbf{F}$ and S. ∎

We can extend the proof to general surfaces for which lines parallel to the coordinate axes intersect the surfaces in more than two points. Most volumes V bounded by surfaces S can be subdivided into n sub-volumes V_i whose bounding surfaces S_i do satisfy this condition (Figure 6.24).

For each such sub-volume, the divergence theorem is now known to apply:

$$\int\int_{S_i}\mathbf{F}\cdot\mathbf{n}\,dS = \int\int\int_{V_i}\nabla\cdot\mathbf{F}\,dV, \quad i = 1,\ldots,n.$$

If these n equations are then added together, we have

$$\sum_{i=1}^n \int\int_{S_i}\mathbf{F}\cdot\mathbf{n}\,dS = \sum_{i=1}^n \int\int\int_{V_i}\nabla\cdot\mathbf{F}\,dV.$$

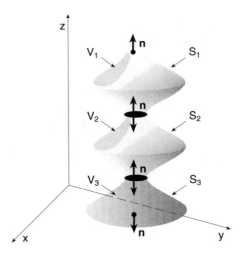

FIGURE 6.25
The closed region for Example 6.12

The right side is the triple integral of $\nabla \cdot \mathbf{F}$ over V because the V_i constitute V.

Figure 6.24 illustrates that when surface integrals over the S_i are added, contributions from interior surfaces cancel out in pairs, and the remaining surface integrals add to give the surface integral of $\mathbf{F} \cdot \mathbf{n}$ over S. The proof for more general surfaces that cannot be subdivided into a finite number of sub-surfaces of this type is left for more advanced texts.

Example 6.12
Let D be the region bounded by the hemisphere $x^2 + y^2 + (z-1)^2 = 9$, $1 \le z \le 4$, and the plane $z = 1$. Verify the divergence theorem if $\mathbf{F} = x\mathbf{i} + y\mathbf{j} + (z-1)\mathbf{k}$. ⬜

SOLUTION   The closed region is shown in Figure 6.25.
The Triple Integral: Because $\mathbf{F} = x\mathbf{i} + y\mathbf{j} + (z-1)\mathbf{k}$, we see that $\nabla \cdot \mathbf{F} = 3$. Hence,

$$\int \int \int_D \nabla \cdot \mathbf{F} \, dV = \int \int \int_D 3 \, dV = 3 \int \int \int_D dV$$

$$= 3 \left(\frac{2}{3} \pi (3)^3 \right) = 54\pi. \tag{6.15}$$

In the last calculation, we have used the fact that $\int \int \int_D dV$ gives the volume of the hemisphere.

The Surface Integral: We write

$$\int\int_S = \int\int_{S_1} + \int\int_{S_2},$$

where S_1 is the hemisphere and S_2 is the plane $z = 1$. If S_1 is a level surface of $g(x, y, z) = x^2 + y^2 + (z - 1)^2$, then a unit outer normal is

$$\mathbf{n} = \frac{\nabla g}{|\nabla g|} = \frac{x\mathbf{i} + y\mathbf{j} + (z - 1)\mathbf{k}}{\sqrt{x^2 + y^2 + z^2}} = \frac{x}{3}\mathbf{i} + \frac{y}{3}\mathbf{j} + \frac{(z - 1)}{3}\mathbf{k}.$$

Now

$$\mathbf{F} \cdot \mathbf{n} = \frac{x^2}{3} + \frac{y^2}{3} + \frac{(z - 1)^2}{3} = 3$$

and so

$$\int\int_{S_1} (\mathbf{F} \cdot \mathbf{n})\, dS = \int\int_R (3) \left(\frac{3}{\sqrt{9 - x^2 - y^2}}\, dA \right).$$

and using polar coordinates,

$$\int\int_R (3) \left(\frac{3}{\sqrt{9 - x^2 - y^2}}\, dA \right) = 9 \int_0^{2\pi} \int_0^3 (9 - r^2)^{-1/2}\, r\, dr\, d\theta$$
$$= 54\pi.$$

On S_2, we take $\mathbf{n} = -\mathbf{k}$ so that $\mathbf{F} \cdot \mathbf{n} = -z + 1$. But because $z = 1$,

$$\int\int_{S_2} (-z + 1)\, dS = 0.$$

Hence, we see that

$$\int\int_R (\mathbf{F} \cdot \mathbf{n})\, dS = 54\pi + 0 = 54\pi,$$

which agrees with (6.16).   ∎

Example 6.13

If $\mathbf{F} = xy\mathbf{i} + y^2z\mathbf{j} + z^3\mathbf{k}$, evaluate $\int\int_S (\mathbf{F} \cdot \mathbf{n})\, dS$, where S is the unit cube defined by $0 \le x \le 1$, $0 \le y \le 1$, $0 \le z \le 1$ (see Figure 6.26)   ☐

SOLUTION   Rather than evaluate six surface integrals, we apply the divergence theorem. Then we need $\nabla \cdot \mathbf{F} = y + 2xy + 3z^2$, so that

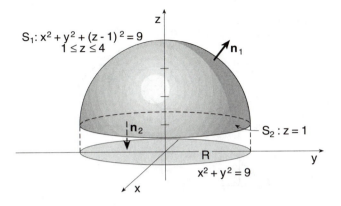

FIGURE 6.26
The unit cube S for Example 6.13

$$\int\!\!\int_S (\mathbf{F} \cdot \mathbf{n})\, dS = \int\!\!\int\!\!\int_D (y + 2yz + 3z^2)\, dV$$
$$= \int_0^1 \int_0^1 \int_0^1 (y + 2yz + 3z^2)\, dx\, dy\, dz$$
$$= \int_0^1 \int_0^1 (y + 2yz + 3z^2)\, dy\, dz$$
$$= \int_0^1 \left(\frac{y^2}{2} + y^2 z + 3yz^2\right)\bigg|_0^1 dz$$
$$= \int_0^1 \left(\frac{1}{2} + z + 3z^2\right) dz$$
$$= \left(\frac{1}{2}z + \frac{z^2}{2} + z^3\right)\bigg|_0^1 = 2.$$

I

Example 6.14
Use the divergence theorem to evaluate the surface integral of Example 6.7.
⬚

SOLUTION   By the divergence theorem (see Figure 6.14), we have

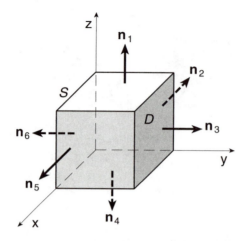

FIGURE 6.27
Volume defined by surfaces given in Example 6.15

$$\iint_S \mathbf{F} \cdot \mathbf{n} \, dS = \iiint_V \nabla \cdot \mathbf{F} \, dV$$
$$= \iiint_V (1 + 1 + 1) \, dV$$
$$= 3 \iiint_V dV$$
$$= 3(\text{volume of } V)$$
$$= 3(4\pi)(2)$$
$$= 24\pi.$$

∎

Example 6.15
Evaluate

$$\iint_S (x^3\mathbf{i} + y^3\mathbf{j} + z^3\mathbf{k}) \cdot \mathbf{n} \, dS$$

where $\mathbf{n}$ is the unit inner normal to the surface bounding the volume defined by the surfaces $y = x$, $z = 0$, $z = 2$, $y = 0$, $x = 4$. ⬜

SOLUTION   By the divergence theorem (See Figure 6.27), we have

$$\int\int_S (x^3\mathbf{i} + y^3\mathbf{j} + z^3\mathbf{k}) \cdot \mathbf{n}\, dS = -\int\int\int_V \nabla \cdot (x^3\mathbf{i} + y^3\mathbf{j} + z^3\mathbf{k})\, dV$$

$$= -\int\int\int_V (3x^3 + 3y^3 + 3z^3)\, dV$$

$$= -3\int_0^4 \int_0^x \int_0^2 (x^2 + y^2 + z^2)\, dz\, dy\, dx$$

$$= -3\int_0^4 \int_0^x \left\{ (x^2 + y^2)z + \frac{z^3}{3} \right\}_0^2 dy\, dx$$

$$= -2\int_2^4 \int_0^x \left\{ 3(x^2 + y^2) + 4 \right\}\, dy\, dx$$

$$= -2\int_0^4 \left\{ y^3 + (3x^2 + 4)y \right\}_0^x\, dx$$

$$= -2\int_0^4 (4x^3 + 4x)\, dx = -2\left(x^4 + 2x^2 \right)\big|_0^4$$

$$= -576.$$

∎

REMARK 6.2   Various important theorems stem from the divergence theorem. For instance, if u and v are two sufficiently differentiable scalar point functions, and if we let

$$\mathbf{F} = u\nabla v,$$

then, by application of the vector formulae in Chapter 4, we have

$$\nabla \cdot \mathbf{F} = \nabla \cdot (u\nabla v)$$
$$= u\nabla \cdot \nabla v + \nabla u \cdot \nabla v$$
$$= \nabla u \cdot \nabla v + u\nabla^2 v.$$

Hence, applying the divergence theorem to the vector $\mathbf{F} = u\nabla v$, we have

$$\int\int_V\int \left(\nabla u \cdot \nabla v + u\nabla^2 v \right)\, dV = \int\int_S \mathbf{n} \cdot u\nabla v\, dS. \qquad (6.16)$$

Similarly, if we interchange the roles of u and v in (6.17), we obtain

$$\int\int_V\int \left(\nabla v \cdot \nabla u + v\nabla^2 u \right)\, dV = \int\int_S \mathbf{n} \cdot v\nabla u\, dS. \qquad (6.17)$$

Finally, if we subtract (6.18) from (6.17), we obtain Green's Theorem

$$\int\int_V\int \left(u\nabla^2 v - v\nabla^2 u\right)\, dV = \int\int_S \mathbf{n}\cdot(u\nabla v - v\nabla u)\, dS \qquad (6.18)$$

where $\mathbf{n}$ is the outward normal to the surface S which bounds the volume V. Green's Theorem in the form (6.19), is of great importance in many areas of research. Some specific areas are in the development of the boundary element method, diffraction of water waves by offshore structures and many other physical problems.   ∎

Exercises

1. Evaluate the line integral by two methods: (i) directly and (ii) using Green's Theorem.

   (a) $\oint_C x^2 y dx + xy^3 dy$, C is the square with vertices $(0,0)$, $(1,0)$, and $(0,1)$

   (b) $\oint_C x dx - x^2 y^2 dy$, C is the triangle with vertices $(0,0)$, $(1,1)$, and $(0,1)$

   (c) $\oint_C (x+2y)dx + (x-2y)dy$, C consists of the arc of the parabola $y = x^2$ from $(0,0)$ to $(1,1)$ followed by the line segment from $(1,1)$ to $(0,0)$

   (d) $\oint_C (x^2 + y^2)dx + 2xy dy$, C consists of the arc of the parabola $y = x^2$ from $(0,0)$ to $(2,4)$ and the line segments from $(2,4)$ to $(0,4)$ and from $(0,4)$ to $(0,0)$

2. Use Green's Theorem to evaluate the line integral along the given positively oriented curve.

   (a) $\oint_C xy dx + y^5 dy$, C is the triangle with vertices $(0,0)$, $(2,0)$, and $(2,1)$

   (b) $\oint_C (y+e^{\sqrt{x}})dx + (2x + \cos y^2)dy$, C is the boundary of the region enclosed by the parabolas $y = x^2$ and $x = y^2$

   (c) $\oint_C (y - \tan^{-1} x)dx + (3x + \sin y)dy$, C is the boundary of the region enclosed by the parabola $y = x^2$ and the line $y = 4$

   (d) $\oint_C x^2 dx + y^2 dy$, C is the curve $x^6 + y^6 = 1$

   (e) $\oint_C 2xy dx + x^2 dy$, C is the cardioid $r = 1 + \cos\theta$

   (f) $\oint_C (x^3 - y^3)dx + (x^3 + y^3)dy$, C is the boundary of the region between the circles $x^2 + y^2 = 1$ and $x^2 + y^2 = 9$

   (g) $\int_C \mathbf{F}\cdot d\mathbf{r}$, where $\mathbf{F}(x,y) = (y^2 - x^2 y)\mathbf{i} + xy\mathbf{j}$ and C consists of the circle $x^2 + y^2 = 4$ from $(2,0)$ to $(\sqrt{2},\sqrt{2})$ to $(0,0)$ and from $(0,0)$ to $(2,0)$

(h) $\int_C \mathbf{F} \cdot d\mathbf{r}$, where $\mathbf{F}(x, y) = x^3 y \mathbf{i} + x^4 \mathbf{j}$ and C is the curve $x^4 + y^4 = 1$

3. Use Green's Theorem to find the work done by the force $\mathbf{F}(x, y) = x(x + y)\mathbf{i} + xy^2\mathbf{j}$ in moving a particle from the origin along the x-axis to $(1, 0)$, then along the line segment to $(0, 1)$, and then back to the origin along the y-axis.

4. A particle starts at the point $(-2, 0)$, moves along the x-axis to $(2, 0)$, and then along the semi-circle $y = \sqrt{4 - x^2}$ to the starting point. Use Green's Theorem to find the work done on this particle by the force field $\mathbf{F}(x, y) = (x, x^3 + 3xy^2)$.

5. Use Stokes' Theorem to evaluate $\int\int_S \nabla \times \mathbf{F} \cdot d\mathbf{S}$.

   (a) $\mathbf{F}(x, y, z) = xyz\mathbf{i} + xj + e^{xy} \cos z\mathbf{k}$, S is the hemisphere $x^2 + y^2 + z^2 = 1$, $0 \leq z$, oriented upward

   (b) $\mathbf{F}(x, y, z) = y^2 z\mathbf{i} + xz\mathbf{j} + x^2 y^2 \mathbf{k}$, S is the part of the paraboloid $z = x^2 + y^2$ that lies inside the cylinder $x^2 + y^2 = 1$, oriented upward

   (c) $\mathbf{F}(x, y, z) = yz^3\mathbf{i} + \sin(xyz)\mathbf{j} + x^3\mathbf{k}$, S is the part of the paraboloid $y = 1 - x^2 - z^2$ that lies to the right of the xz-plane, oriented toward the xz-plane

   (d) $\mathbf{F}(x, y, z) = (x + \tan^{-1} yz)\mathbf{i} + y^2 z\mathbf{j} + z\mathbf{k}$, S is the part of the hemisphere $x = \sqrt{9 - y^2 - z^2}$ that lies inside the cylinder $y^2 + z^2 = 4$, oriented in the direction of the positive x-axis

In the next four questions, use Stokes' Theorem to evaluate $\int_C \mathbf{F} \cdot d\mathbf{r}$. In each case, C is oriented counterclockwise as viewed from above.

6. $\mathbf{F}(x, y, z) = xz\mathbf{i} + 2xy\mathbf{j} + 3xy\mathbf{k}$, C is the boundary of the part of the plane $3x + y + z = 3$ in the first octant

7. $\mathbf{F}(x, y, z) = z^2\mathbf{i} + y^2\mathbf{j} + xy\mathbf{k}$, C is the triangle with vertices $(1, 0, 0)$, $(0, 1, 0)$, and $(0, 0, 2)$

8. $\mathbf{F}(x, y, z) = 2z\mathbf{i} + 4x\mathbf{j} + 5y\mathbf{k}$, C is the curve of intersection of the plane $z = x + 4$ and the cylinder $x^2 + y^2 = 4$

9. $\mathbf{F}(x, y, z) = x\mathbf{i} + y\mathbf{j} + (x^2 + y^2)\mathbf{k}$, C is the part of the paraboloid $z = 1 - x^2 - y^2$ in the first octant

10. Use Stokes' Theorem to evaluate $\int_C \mathbf{F} \cdot d\mathbf{r}$, where

$$\mathbf{F}(x, y, z) = x^2 z\mathbf{i} + xy^2\mathbf{j} + z^2\mathbf{k}$$

and C is the curve of intersection of the plane $x + y + z = 1$ and the cylinder $x^2 + y^2 = 9$ oriented counterclockwise as viewed from above.

11. Use Stokes' Theorem to evaluate $\int_C \mathbf{F} \cdot d\mathbf{r}$, where $\mathbf{F}(x, y, z) = x^2 y \mathbf{i} + \frac{1}{3}x^3 \mathbf{j} + xy\mathbf{k}$ and C is the curve of intersection of the hyperbolic paraboloid $z = y^2 - x^2$ and the cylinder $x^2 + y^2 = 1$ oriented counterclockwise as viewed from above.

12. Verify that Stokes' Theorem is true for the given vector field $\mathbf{F}$ and surface S.

(a) $\mathbf{F}(x, y, z) = 3y\mathbf{i} + 4z\mathbf{j} - 6x\mathbf{k}$, S is the part of the paraboloid

$$z = 9 - x^2 - y^2$$

that lies above the xy-plane, oriented upward

(b) $\mathbf{F}(x, y, z) = xy\mathbf{i} + yz\mathbf{j} + xz\mathbf{k}$, S is the hemisphere

$$z = \sqrt{a^2 - x^2 - y^2},$$

oriented upward

(c) $\mathbf{F}(x, y, z) = y\mathbf{i} + z\mathbf{j} + x\mathbf{k}$, S is the part of the plane $x + y + z = 1$ that lies in the first octant, oriented upward

13. Calculate the work done by the force field

$$\mathbf{F}(x, y, z) = (x^x + z^2)\mathbf{i} + (y^y + x^2)\mathbf{j} + (z^z + y^2)\mathbf{k}$$

when a particle moves under its influence around the edge of the part of the sphere $x^2 + y^2 + z^2 = 4$ that lies in the first octant, in a counterclockwise direction as viewed from above.

14. Evaluate $\int_C (y + \sin x)dx + (z^2 + \cos y)dy + x^3 dz$, where C is the curve $\mathbf{r}(t) = (\sin t, \cos t, \sin 2t)$, $0 \le t \le 2\pi$. [*Hint:* Observe that C lies on the surface $z = 2xy$.]

In the next 7 questions, use the Divergence Theorem to calculate the surface integral $\int \int_S \mathbf{F} \cdot d\mathbf{S}$; i.e., calculate the flux of $\mathbf{F}$ across S.

15. $\mathbf{F}(x, y, z) = -xz\mathbf{i} - yz\mathbf{j} + z^2\mathbf{k}$, S is the ellipsoid $x^2/a^2 + y^2/b^2 + z^2/c^2 = 1$

16. $\mathbf{F}(x, y, z) = z \cos y\mathbf{i} + x \sin z\mathbf{j} + xz\mathbf{k}$, S is the surface of the tetrahedron bounded by the planes $x = 0$, $y = 0$, $z = 0$, and $2x + y + z = 2$

17. $\mathbf{F}(x, y, z) = (x + e^{y \tan z})\mathbf{i} + 3xe^{xz}\mathbf{j} + (\cos y - z)\mathbf{k}$, S is the surface with equation $x^4 + y^4 + z^4 = 1$

18. $\mathbf{F}(x, y, z) = x^3\mathbf{i} + y^3\mathbf{j} + z^3\mathbf{k}$, S is the sphere $x^2 + y^2 + z^2 = 1$

19. $\mathbf{F}(x, y, z) = (x^3 + y \sin z)\mathbf{i} + (y^3 + z \sin x)\mathbf{j} + 3z\mathbf{k}$, S is the surface of the solid bounded by the hemispheres $z = \sqrt{4 - x^2 - y^2}$, $z = \sqrt{1 - x^2 - y^2}$ and the plane $z = 0$

20. $\mathbf{F}(x, y, z) = xy^2\mathbf{i} + yz\mathbf{j} + zx^2\mathbf{k}$, S is the surface of the solid that lies between the cylinders $x^2 + y^2 = 1$ and $x^2 + y^2 = 4$ and between the planes $z = 1$ and $z = 3$

21. $\mathbf{F}(x, y, z) = (x^3 + yz)\mathbf{i} + x^2 y\mathbf{j} + xy^2\mathbf{k}$, S is the surface of the solid bounded by the spheres $x^2 + y^2 + z^2 = 4$ and $x^2 + y^2 + z^2 = 9$

22. Use the Divergence Theorem to evaluate $\int\int_S \mathbf{F} \cdot d\mathbf{S}$, where

$$\mathbf{F}(x, y, z) = z^2 x\mathbf{i} + (\frac{1}{3}y^3 + \tan z)\mathbf{j} + (x^2 z + y^2)\mathbf{k}$$

and S is the top half of the sphere $x^2 + y^2 + z^2 = 1$. [*Hint:* Note that S is not a closed surface. First compute integrals over S_1 and S_2, where S_1 is the disk $x^2 + y^2 \leq 1$ oriented downward and $S_2 = S \cup S_1$.]

23. Let $\mathbf{F}(x, y, z) = z \tan^{-1}(y^2)\mathbf{i} + z^2 \ln(x^2 + 1)\mathbf{j} + z\mathbf{k}$. Find the flux of $\mathbf{F}$ across the part of the paraboloid $x^2 + y^2 + z = 2$ that lies above the plane $z = 1$ and is oriented upward.

24. Verify that div $\mathbf{E} = 0$ for the electric field $\mathbf{E}(\mathbf{x}) = \varepsilon Q\mathbf{x}/|\mathbf{x}|^3$.

25. Use the Divergence Theorem to evaluate

$$\int\int_S (2x + 2y + z^2) dS$$

where S is the sphere $x^2 + y^2 + z^2 = 1$.

26. Verify that Stokes' Theorem is true for the vector field $\mathbf{F}(x, y, z) = x^2\mathbf{i} + y^2\mathbf{j} + z^2\mathbf{k}$, where S is the part of the paraboloid $z = 1 - x^2 - y^2$ that lies above the xy-plane, and S has upward orientation.

27. Use Stokes' Theorem to evaluate $\int\int_S \nabla \times \mathbf{F} \cdot d\mathbf{S}$, where $\mathbf{F}(x, y, z) = x^2 yz\mathbf{i} + yz^2\mathbf{j} + z^3 e^{xy}\mathbf{k}$, S is the part of the sphere $x^2 + y^2 + z^2 = 5$ that lies above the plane $z = 1$, and S is oriented upward.

28. Use Stokes' Theorem to evaluate $\int_C \mathbf{F} \cdot d\mathbf{r}$, where $\mathbf{F}(x, y, z) = xy\mathbf{i} + yz\mathbf{j} + zx\mathbf{k}$ and C is the triangle with vertices $(1, 0, 0)$, $(0, 1, 0)$, and $(0, 0, 1)$, oriented counterclockwise as viewed from above.

29. Use the Divergence Theorem to calculate the surface integral $\int\int_S \mathbf{F} \cdot d\mathbf{S}$, where $\mathbf{F}(x, y, z) = x^3\mathbf{i} + y^3\mathbf{j} + z^3\mathbf{k}$ and S is the surface of the solid bounded by the cylinder $x^2 + y^2 = 1$ and the planes $z = 0$ and $z = 2$.

30. Verify that the Divergence Theorem is true for the vector field

$$\mathbf{F}(x, y, z) = x\mathbf{i} + y\mathbf{j} + z\mathbf{k},$$

where S is the unit ball $x^2 + y^2 + z^2 \leq 1$.

31. Compute the outward flux of

$$\mathbf{F}(x, y, z) = \frac{x\mathbf{i} + y\mathbf{j} + z\mathbf{k}}{(x^2 + y^2 + z^2)^{3/2}}$$

through the ellipsoid $4x + 9y^2 + 6z^2 = 36$.

Chapter 7

Applications

7.1   Introduction

In this chapter we will consider a few applications concerning fluid flow in compressible and incompressible fluids, Poisson's equation, electromagnetic theory, Maxwell's wave equations and nonlinear ocean wave equations. A small section on graphical simulation of a vector field has also been included. It is well known that one of the most important uses of vector analysis is in the concise formulation of physical laws and the derivation of other results from those laws.

We develop the mathematical formulae governing these problems using vector analysis in concise form and then we illustrate these theories with the practical problems cited above. We have selected elementary problems to demonstrate the usefulness of vector analysis.

Section 7.10 and 7.11 deal with nonlinear ocean waves with resonance conditions. This material is very advanced and requires an advanced knowledge of the physical situation of the problem. The section will be useful to those working on the extraction of energy from nonlinear ocean waves. The last section demonstrates computer simulation of a vector field with the help of the computer package *Mathematica 3.0*.

7.2   Acceleration Vector

The characteristics of a fluid as it flows past any fixed point in space are a function of the position vector $\mathbf{R} = x\mathbf{i} + y\mathbf{j} + z\mathbf{k}$ and the time variable t. As a particular instance, the velocity vector $\mathbf{V}$, at a point will be a function of the independent variables $\mathbf{R}$ and t. When obtaining the general equations of motion, the general acceleration vector of a moving fluid particle

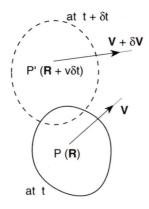

FIGURE 7.1
Acceleration vector

corresponding, to a fixed set of orthogonal axes, is established and then applied to a moving set of orthogonal axes.

Suppose that we let the fluid particle in space occupy the position P at time t. This is equivalent to letting the position vector P be given by the vector $\mathbf{R}$. We can also let the resultant velocity at P be given by $\mathbf{V} = \mathbf{V}(r, t)$. If, after an interval of time δt, the fluid particle moves to the new position P', we can write the new position vector as $\mathbf{R} + \mathbf{V}\delta t$. The velocity vector of this new position vector at P' can be written as $\mathbf{V} + \delta\mathbf{V} = \mathbf{V}(\mathbf{R} + \mathbf{V}\delta t, t + \delta t)$ (see Figure 7.1). This leads us to the following expression:

$$
\begin{aligned}
\delta\mathbf{V} &= \mathbf{V}(\mathbf{R} + \mathbf{V}\delta t, t + \delta t) - \mathbf{V}(\mathbf{R}, t) \\
&= \mathbf{V}(\mathbf{R} + \mathbf{V}\delta t, t + \delta t) - \mathbf{V}(\mathbf{R}, t + \delta t) \\
&\quad + \mathbf{V}(\mathbf{R}, t + \delta) - \mathbf{V}(\mathbf{R}, t).
\end{aligned}
$$

By Taylor's theorem, the preceding equation becomes

$$
\begin{aligned}
\mathbf{V}(\mathbf{R} + \mathbf{V}\delta t, t + \delta t) &- \mathbf{V}(\mathbf{R}, t + \delta t) \\
&= (\mathbf{V}\delta t \cdot \mathrm{grad})\mathbf{V}(\mathbf{R}, t + \delta t) + O(\delta t^2),
\end{aligned}
$$

$$
\text{and } \ \mathbf{V}(\mathbf{R}, t + \delta t) - \mathbf{V}(\mathbf{R}, t) = \left(\frac{\partial \mathbf{V}(\mathbf{R}, t)}{\partial t}\right)\delta t + O(\delta t^2).
$$

By addition, we get

$$
\delta\mathbf{V} = \left(\frac{\partial \mathbf{V}(\mathbf{R}, t)}{\partial t} + (\mathbf{V} \cdot \mathrm{grad})\mathbf{V}(\mathbf{R}, t + \delta t)\right)\delta t + O(\delta t^2).
$$

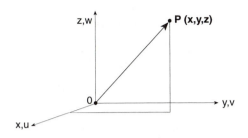

FIGURE 7.2
Cartesian coordinate axes

Now, if we divide throughout by δt and let $\delta t \to 0$,

$$\lim_{\delta t \to 0} \left(\frac{\delta \mathbf{V}}{\delta t} \right) = \left(\frac{\partial \mathbf{V}}{\partial t} \right) + (\mathbf{V} \cdot \text{grad})\mathbf{V}.$$

Then from the definition of differentiation, the total rate of change $\frac{D\mathbf{V}}{Dt}$ is defined as the rate of change of $\mathbf{V}$ following a particle of fluid

$$\frac{D\mathbf{V}}{Dt} = \frac{\partial \mathbf{V}}{\partial t} + (\mathbf{V} \cdot \text{grad})\mathbf{V}, \tag{7.1}$$

which is nothing but the sum of the local and convective rates of change of $\mathbf{V}$. In this expression, (equation (7.1)), the operator $(D/Dt) = (\partial/\partial t) + (\mathbf{V} \cdot \text{grad})$ is the total differentiation following the motion of the fluid. The interested reader is referred to the work by Lighthill[7] that gives a very thorough description of this operator from the physical point of view.

The operator D/Dt is given below for the three commonly used orthogonal coordinate axes.

1. *Rectangular Cartesian coordinates.*
   The velocity vector is given by $\mathbf{V} = (u, v, w)$ (see Figure 7.2). Elementary lengths are $(\delta x, \delta y, \delta z)$,

$$\text{grad} = \nabla = [(\partial/\partial x), (\partial/\partial y), (\partial/\partial z)].$$

   Hence, we have the following:

$$\frac{D}{Dt} = \frac{\partial}{\partial t} + u \frac{\partial}{\partial x} + v \frac{\partial}{\partial y} + w \frac{\partial}{\partial z} \cdots$$

2. *Cylindrical polar coordinates*
   The well-known relationship between Cartesian and cylindrical polar coordinates is written as $x = r \cos \theta$, $y = r \sin \theta$, and $z = z$ (see Figure

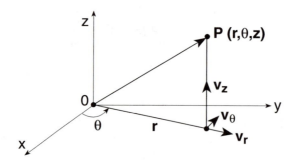

FIGURE 7.3
Cylindrical polar coordinate axes

7.3). The elementary lengths in this case are $(\delta r, r\delta\theta, \delta z)$. The velocity vector $\mathbf{V} = (u_r, v_\theta, v_z)$ and

$$\nabla = [(\partial/\partial r), (\partial/r\partial\theta), (\partial/\partial z)].$$

This leads us to

$$\frac{D}{Dt} = \frac{\partial}{\partial t} + v_r\frac{\partial}{\partial r} + \frac{v_\theta}{r}\frac{\partial}{\partial\theta} + v_z\frac{\partial}{\partial z}.$$

3. *Spherical polar coordinates*

The relationship between Cartesian and spherical polar coordinates can be written as $x = (r\sin\phi)\cos\theta$, $y = (r\sin\phi)\sin\theta$, and $z = r\cos\phi$ (see Figure 7.4). The corresponding elementary lengths are given by $(\delta r, r\delta\phi, r\sin\phi\delta\theta)$, the velocity vector $\mathbf{V} = (v_r, v_\phi, v_\theta)$ and $\nabla = [(\partial/\partial r), (\partial/r\partial\phi), (1/r\sin\phi)(\partial/\partial\theta)]$. Therefore, we obtain

$$\frac{D}{Dt} = \frac{\partial}{\partial t} + v_r\frac{\partial}{\partial r} + \frac{v_\phi}{r}\frac{\partial}{\partial\phi} + \frac{v_\theta}{r\sin\phi}\frac{\partial}{\partial\theta}.$$

For a Cartesian coordinate system, the components of the velocity vector $\mathbf{V}$ would (at all times and at all points in space) be parallel to the Cartesian axes. However, when we the velocity vector $\mathbf{V}$ is considered under a polar coordinate system, there is a rotation of the axes. At time t the position of P' will be $\mathbf{R}(r, \phi, \theta)$ where the velocity is given by $\mathbf{V}(v_r, v_\phi, v_\theta)$.

After the time interval δt, the fluid will have moved to position P''. At this position, the components of velocity, v_r, v_ϕ, and v_θ will now be parallel to the axes r', ϕ' and θ'. Now, if we refer to Curle[2], we can define the acceleration of the fluid particle in the moving frame of axes as

$$\text{acceleration} = \frac{D\mathbf{V}}{Dt} + [\mathbf{\Omega} \times \mathbf{V}],$$

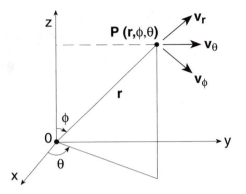

FIGURE 7.4
Spherical polar coordinate axes

where $\boldsymbol{\Omega}$ is the angular velocity vector of axes.

In the spherical polar coordinate system, the angular velocity vector is given by

$$\boldsymbol{\Omega} = [(v_\theta/r)\cot\phi, -(v_\theta/r), (v_\phi/r)]$$

and the linear velocity vector can be written as $\mathbf{V} = (v_r, v_\phi, v_\theta)$. This will then lead to

$$[\boldsymbol{\Omega}\times\mathbf{V}] = \begin{vmatrix} \mathbf{i} & \mathbf{j} & \mathbf{k} \\ \frac{v_\theta}{r}\cot\phi & -\frac{v_\theta}{r} & \frac{v_\phi}{r} \\ v_r & v_\phi & v_\theta \end{vmatrix}.$$

Expansion of this vector product leads to the component

$$\left(-\frac{v_\theta^2 + v_\phi^2}{r}, \frac{v_r v_\phi}{r} - \frac{v_\theta^2}{r}\cot\phi, \frac{v_\theta v_\phi}{r}\cot\phi + \frac{v_r v_\theta}{r}\right).$$

It follows that the components of the acceleration vector in spherical polar coordinates can be written in the following manner

$$\frac{Dv_r}{Dt} - \frac{v_\theta^2 + v_\phi^2}{r}, \frac{Dv_\phi}{Dt} + \frac{v_r v_\phi}{r} - \frac{v_\theta^2}{r}\cot\phi, \frac{Dv_\theta}{Dt} + \frac{v_r v_\theta}{r} + \frac{v_\theta v_\phi}{r}\cot\phi.$$

In the cylindrical polar coordinate axes, the angular velocity components are $\boldsymbol{\Omega} = [0, 0, (v_\theta/r)]$.

Note that this result can be obtained from the angular velocity vector $\boldsymbol{\Omega}$ with respect to the spherical polar coordinate system by changing ϕ to θ and then letting $v_\phi = 0$. The linear velocity vector is $\mathbf{V} = (v_r, v_\theta, v_z)$. This leads to

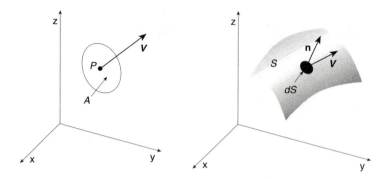

FIGURE 7.5
(a) Gas flowing through P (b) through surface S

$$[\mathbf{\Omega} \times \mathbf{V}] = \begin{vmatrix} \mathbf{i} & \mathbf{j} & \mathbf{k} \\ 0 & 0 & \frac{v_\theta}{r} \\ v_r & v_\theta & v_z \end{vmatrix} = \left(-\frac{v_\theta^2}{r}, \frac{v_r v_\theta}{r}, 0 \right).$$

Thus, the components of the acceleration vector in a cylindrical polar coordinate system can be written as

$$\left(\frac{Dv_r}{Dt} - \frac{v_\theta^2}{r}, \frac{Dv_\theta}{Dt} + \frac{v_r v_\theta}{r}, \frac{Dv_z}{Dt} \right).$$

7.3   Continuity Equation of Fluid Flow

In many branches of engineering and physics we encounter the concepts of **flux** and **circulation**. We will explore the relationships between flux and divergence and between circulation and curl.

Consider a gas flowing through a region D of space. At time t and point $P(x, y, z)$ in D, gas flows through P with velocity $\mathbf{V}(x, y, z, t)$. If A is a unit area around P perpendicular to $\mathbf{V}$ (Figure 7.5(a)) and $\rho(x, y, z, t)$ is density of the gas at P, then the amount of gas crossing A per unit area is $\rho \mathbf{V}$. At every point P in D, the vector $\rho \mathbf{V}$ is such that its direction $\mathbf{V}$ gives velocity of gas flow, and its magnitude $\rho |\mathbf{V}|$ describes the mass of gas flowing in that direction per unit time.

Consider some surface S in D (Figure 7.6). If $\mathbf{n}$ is a unit normal to S, then $\rho \mathbf{V} \cdot \mathbf{n}$ is the component of $\rho \mathbf{V}$ normal to the surface S. If dS is an element of area on S, then $\rho \mathbf{V} \cdot \mathbf{n} \, dS$ describes the mass of gas flowing through dS per unit time. Consequently,

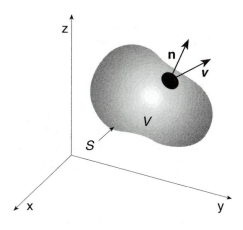

FIGURE 7.6
Closed surface S with unit outer normal n

$$\int\int_S \rho \mathbf{V} \cdot \mathbf{n}\, dS$$

is the mass of gas flowing through S per unit time. This quantity, as mentioned before, is known as the flux for surface S. If S is a closed surface (Figure 7.6) and **n** is the unit outward normal to S then

$$\int\int_S \rho \mathbf{V} \cdot \mathbf{n}\, dS$$

is the mass of gas flowing out of surface S per unit time. If this flux (for closed S) is positive then there is a net outward flow of gas through (i.e., more gas is leaving the volume bounded by S than is entering); if the flux is negative, the net flow is inward.

If we now apply the divergence theorem to the flux integral over the closed surface S, we have

$$\int\int_S \rho \mathbf{V} \cdot \mathbf{n}\, dS = \int\int\int_V \nabla \cdot (\rho \mathbf{V})\, dV. \tag{7.2}$$

Now the flux (the integral on the left of this equation) is the mass of gas per unit time leaving S. In order for the right side to represent the same quantity, $\nabla \cdot (\rho \mathbf{V})$ must be interpreted as the mass of gas leaving unit volume per unit time, because then $\nabla \cdot (\rho \mathbf{V})\, dV$ represents the mass per unit time leaving dV, and the triple integral is the mass per unit time leaving V. We have thus obtained an interpretation of the divergence of $\rho \mathbf{V}$. The divergence of $\rho \mathbf{V}$ is the flux per unit volume per unit time at a point: the mass of gas leaving unit volume in unit time. We can use this idea to derive the *equation of continuity* for fluid flow. The triple integral

$$\int \int \int_V \frac{\partial \rho}{\partial t} \, dV = \frac{\partial}{\partial t} \int \int \int_V \rho \, dV$$

measures the time rate of change of mass in a volume V. If this triple integral is positive, then there is a net inward flow of mass; if it is negative, the net flow is outward. We conclude that this triple integral must be the negative of the flux for the volume V, i.e.,

$$\int \int \int_V \frac{\partial \rho}{\partial t} \, dV = - \int \int \int_V \nabla \cdot (\rho \mathbf{V}) \, dV,$$

or

$$\int \int \int_V \left(\nabla \cdot (\rho \mathbf{V}) + \frac{\partial \rho}{\partial t} \right) dV = 0.$$

If $\nabla \cdot (\rho \mathbf{V})$ and $\partial \rho / \partial t$ are continuous functions, then this equation can hold for arbitrary volume V only if

$$\nabla \cdot (\rho \mathbf{V}) + \frac{\partial \rho}{\partial t} = 0. \tag{7.3}$$

This is called the **equation of continuity** and expresses conservation of mass. It is basic to all fluid flow. If C is a closed curve in the flow region D, then the **circulation** of the flow for the curve C is defined by

$$\Gamma = \oint_C \mathbf{V} \cdot d\mathbf{R}. \tag{7.4}$$

To obtain an intuitive feeling for Γ, we consider two very simple two-dimensional flows. First, suppose $\mathbf{V} = x\mathbf{i} + y\mathbf{j}$, so that all particles of gas flow along radial lines directed away from the origin as shown in Figure 7.7. In this case, the line integral defining Γ is independent of path and $\Gamma = 0$ for any curve whatsoever.

Second, suppose $\mathbf{V} = -y\mathbf{i} + x\mathbf{j}$, so that all particles of the gas flow counterclockwise around circles centered at the origin (Figure 7.7). In this case, Γ does not generally vanish. In particular, if C is the circle $x^2 + y^2 = r^2$, then $\mathbf{V}$ and $d\mathbf{R}$ are parallel, and

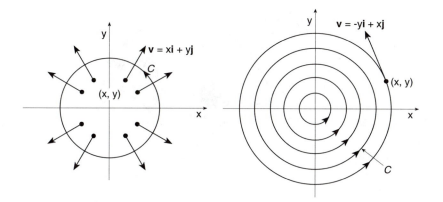

FIGURE 7.7
Two-dimensional gas flows

$$\Gamma = \oint_C \mathbf{V} \cdot d\mathbf{R}$$

$$= \oint_C \mathbf{V} \cdot \frac{d\mathbf{R}}{dS} dS, \text{ this can be written as}$$

$$= \oint_C \mathbf{V} \cdot \frac{(d\mathbf{R}/dt)}{(dS/dt)} \, dS, \text{ which in turn becomes}$$

$$= \oint_C \frac{\mathbf{V} \cdot \mathbf{V}}{|\mathbf{V}|} \, dS$$

$$= \oint_C |\mathbf{V}| \, dS \text{ which leads to}$$

$$= \oint_C \sqrt{x^2 + y^2} dS = \oint_C r \, dS$$

$$= 2\pi r^2.$$

These two flow patterns indicate perhaps that circulation is a measure of the tendency for the flow to be circulatory. If we apply Stokes' theorem to the circulation integral for the closed curve C, we have

$$\oint_C \mathbf{V} \cdot d\mathbf{R} = \int\int_S (\nabla \times \mathbf{V}) \cdot \mathbf{n} \, dS \qquad (7.5)$$

where S is any surface in the flow with boundary C. If the right side of this equation is to represent the circulation for C also, then $(\nabla \times \mathbf{V}) \cdot \mathbf{n} \, dS$ must be interpreted as the circulation for the curve bounding dS (or simply dS itself). Then the addition process of the surface integral (Figure 7.8)

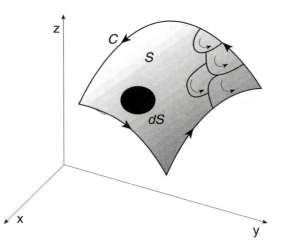

FIGURE 7.8
Surface integral giving circulation around C

gives the circulation around C, the circulation around all internal boundaries cancelling. But, if $(\nabla \times \mathbf{V}) \cdot \mathbf{n}\, dS$ is the circulation for dS, then it follows that $(\nabla \times \mathbf{V}) \cdot$ must be the circulation for unit area perpendicular to $\mathbf{n}$. Thus, $\nabla \times \mathbf{V}$ describes the circulatory nature of the flow $\mathbf{V}$; its component $(\nabla \times \mathbf{V}) \cdot \mathbf{n}$ in any direction describes the circulation for unit area perpendicular to $\mathbf{n}$.

If we examine the continuity equation more closely, we can observe some additional physical applications

$$\nabla \cdot (\rho \mathbf{V}) + \frac{\partial \rho}{\partial t} = 0. \tag{7.6}$$

For the special case of an incompressible fluid, $\partial \rho / \partial t = 0$ so that

$$\nabla \cdot (\rho \mathbf{V}) = \rho \nabla \cdot \mathbf{V} = 0 \text{ or } \nabla \cdot \mathbf{V} = 0.$$

In other words, the velocity vector of an incompressible fluid is one whose divergence vanishes everywhere within a region. The technical term for such vector fields is **solenoidal**.

If the fluid is irrotational, then the curl of the velocity vector must be zero, i.e.,

$$\nabla \times \mathbf{V} = \mathbf{0}.$$

which implies that $\mathbf{V}$ must be the gradient of a scalar function. If $\phi(x, y, z)$ is the scalar function that is known as the velocity potential, then

$$\mathbf{V} = \nabla \phi.$$

If the fluid is incompressible, then $\nabla \cdot \mathbf{V} = 0$ and therefore

$$\nabla \cdot \nabla \phi = 0$$
$$\text{or } \nabla^2 \phi = 0.$$

This equation is known as Laplace's equation, which is satisfied by the velocity potential ϕ.

REMARK 7.1   Let m denote any general property of a particle of the fluid, such as its pressure, density, etc., by the function $Q(x, y, z, t)$. Thus, by $\frac{\partial Q}{\partial t}$ is meant the variation of Q at a particular point in space as a function of time t. If we take the total differential $Q(x, y, z, t)$ with respect to time t, we obtain

$$\frac{dQ}{dt} = \frac{\partial Q}{\partial t} + \frac{\partial Q}{\partial x}\frac{dx}{dt} + \frac{\partial Q}{\partial y}\frac{dy}{dt} + \frac{\partial Q}{\partial z}\frac{dz}{dt}. \tag{7.7}$$

The quantity $\frac{dQ}{dt}$ is called the total derivative or material derivative of Q with respect to time when we fix our attention on the same particle of fluid. Equation (7.7) can very easily be put into vector form as follows

$$\frac{dQ}{dt} = \frac{\partial Q}{\partial t} + \nabla Q \cdot \frac{d\mathbf{R}}{dt}$$
$$= \frac{\partial Q}{\partial t} + \mathbf{V} \cdot \nabla Q \tag{7.8}$$

where

$$\mathbf{V} = \frac{d\mathbf{R}}{dt} = \mathbf{i}\frac{dx}{dt} + \mathbf{j}\frac{dy}{dt} + \mathbf{k}\frac{dz}{dt}$$
$$\text{and } \nabla Q = \mathbf{i}\frac{\partial Q}{\partial x} + \mathbf{j}\frac{\partial Q}{\partial y} + \mathbf{k}\frac{\partial Q}{\partial z}.$$

7.3.1   Euler's Equation of motion

To develop the equation of motion of a nonviscous fluid, we consider the forces acting on an element of fluid with elementary volume $dx\,dy\,dz$ such that if $\rho(x, y, z, t)$ is the density of fluid, then its mass is $\rho\,dx\,dydz$. Thus, if $P(x, y, z, t)$ is the pressure acting on the faces of the elementary volume of the fluid, then there will be a force in the x-direction on the element of the fluid under consideration of the magnitude

$$Pdydz - (P + \frac{\partial P}{\partial x} dx)dydz = -\frac{\partial P}{\partial x} dxdydz. \qquad (7.9)$$

Let us also consider the action of an external force $\mathbf{F} = (F_x, F_y, F_z)$ acting on a unit mass of fluid. The external force acting on the element of fluid in the x-direction is $F_x(\rho\, dx\, dy\, dz)$. The acceleration of the element in the x-direction is $\frac{du}{dt}$, where u is the x component of the velocity vector $\mathbf{V} = (u, v, w)$. Hence, by Newton's second law of motion, we have

$$\left(\frac{Du}{Dt}\right)(\rho dxdydz) = F_x(\rho dxdydz) - \left(\frac{\partial P}{\partial x}\right)dxdydz.$$

Dividing throughout by $\rho\, dx\, dy\, dz$, we obtain

$$\frac{Du}{Dt} = F_x - \frac{1}{\rho}\frac{\partial P}{\partial x}.$$

By this reasoning, and considering the y and z direction, we obtain Euler's equations of motion:

$$\left.\begin{array}{l}\frac{Du}{dt} = F_x - \frac{1}{\rho}\frac{\partial P}{\partial x} \\ \frac{Dv}{dt} = F_y - \frac{1}{\rho}\frac{\partial P}{\partial y} \\ \frac{Dw}{dt} = F_z - \frac{1}{\rho}\frac{\partial P}{\partial z}.\end{array}\right\}$$

The three scalar equations may be combined into the single vector equation

$$\frac{D\mathbf{V}}{Dt} = \mathbf{F} - \frac{1}{\rho}\nabla P. \qquad (7.11)$$

But

$$\frac{D\mathbf{V}}{Dt} = \frac{\partial \mathbf{V}}{\partial t} + (\mathbf{V} \cdot \nabla)\mathbf{V}. \qquad (7.12)$$

Therefore (7.11) can be written as

$$\frac{\partial \mathbf{V}}{\partial t} + (\mathbf{V} \cdot \nabla)\mathbf{V} = \mathbf{F} - \frac{1}{\rho}\nabla P. \qquad (7.13)$$

Now by the vector identity,

$$\nabla(\mathbf{U} \cdot \mathbf{V}) = (\mathbf{U} \cdot \nabla)\mathbf{V} + (\mathbf{V} \cdot \nabla)\mathbf{U} + \mathbf{U} \times (\nabla \times \mathbf{V})$$
$$+ \mathbf{V} \times (\nabla \times \mathbf{U})$$

if we let $\mathbf{U} = \mathbf{V}$, we get

$$(\mathbf{V} \cdot \nabla)\mathbf{V} = \frac{1}{2}\nabla q^2 - \mathbf{V} \times (\nabla \times \mathbf{V}), \qquad (7.14)$$

where $q^2 = \mathbf{V} \cdot \mathbf{V} = u^2 + v^2 + w^2$. Hence, by using this relationship in (7.13) we may write Euler's equation of motion in the following form:

$$\frac{\partial \mathbf{V}}{\partial t} + \frac{1}{2}\nabla q^2 - \mathbf{V} \times (\nabla \times \mathbf{V}) = \mathbf{F} - \frac{1}{\rho}\nabla P. \qquad (7.15)$$

If the external $\mathbf{F}$ is conservative, it has a potential V and we have

$$\mathbf{F} = -\nabla V.$$

If the motion of the fluid is irrotational, $\nabla \times \mathbf{V} = \mathbf{0}$ and $\mathbf{V} = \nabla\phi$, where ϕ is a scalar potential. In this case, equation (7.15) can be written as

$$\nabla\left(\frac{\partial \phi}{\partial t}\right) + \frac{1}{2}\nabla q^2 + \nabla V + \frac{1}{\rho}\nabla P = 0 \qquad (7.16)$$

or

$$\nabla W = 0 \qquad (7.17)$$

where

$$W = \frac{\partial \phi}{\partial t} + \frac{1}{2}q^2 + V + \frac{P}{\rho}.$$

Now, if $d\mathbf{R} = \mathbf{i}dx + \mathbf{j}dy + \mathbf{k}dz$ denotes an arbitrary point in the fluid at any instant, we have by the scalar product of ∇W and $d\mathbf{R}$

$$\nabla W \cdot d\mathbf{R} = \frac{\partial W}{\partial x}dx + \frac{\partial W}{\partial y}dy + \frac{\partial W}{\partial z}dz = dW$$

such that from (7.17)

$$dW = 0$$

and, consequently, after integration,

$$W = c(t), \qquad (7.18)$$

where $c(t)$ is an arbitrary function of time, because we have integrated along an arbitrary path in the fluid at any instant. Thus we have

$$W = \frac{\partial \phi}{\partial t} + \frac{1}{2}q^2 + V + \int \frac{dP}{\rho} = c(t). \qquad (7.19)$$

Equation (7.19) is known as Bernoulli's equation. In the special case where ρ is a constant and the motion is steady so that $\frac{\partial \phi}{\partial t} = 0$, this equation becomes

$$\frac{q^2}{2} + V + \frac{P}{\rho} = c \qquad (7.20)$$

where c is now a constant. This equation states that per unit mass of fluid, the sum of the kinetic energy, the potential energy V, and the pressure energy P/ρ has a constant volume c for all points of the fluid.

REMARK 7.2　　In aerodynamics the variations of V are usually so small that V can be neglected. Therefore, we can write

$$\rho \frac{q^2}{2} + P = P_0 \tag{7.21}$$

where P_0 is the pressure when the fluid is at rest. It can easily be seen that the pressure diminishes as the velocity increases in the form

$$P = P_0 - \rho \frac{q^2}{2}.$$

From equation (7.21), the speed of the object can be determined from the following equation

$$q = \sqrt{\frac{2(P_0 - P)}{\rho}}. \tag{7.22}$$

The physical interpretation is that airplanes are equipped with instruments for measuring P and P_0 so that the speed of the machine relative to the air can be determined by equation (7.22). ∎

7.4　Continuity Equation and Heat Conduction

These ideas are readily extended to the analysis of heat conduction. The rate at which heat crosses a unit area normal to the direction of flow depends upon the thermal conductivity κ of the material present and the temperature gradient ∇u. Thus, we have

$$\int\int_S (-\kappa \nabla u) \cdot \mathbf{n} \, dS,$$

which gives the rate of heat flow across a closed surface S. But if ρ and c denote the density and specific hear capacity, respectively, then

$$\frac{\partial}{\partial t} \int \int \int_V c\,\rho u \, dV$$

must be the rate of increase of the heat contained in the volume V enclosed by S. Then by continuity, we must have

$$\int\int_S (\kappa \nabla u) \cdot \mathbf{n} \, dS = \frac{\partial}{\partial t} \int\int\int_V c\rho u \, dV$$

$$= \int\int\int_V c\rho \frac{\partial u}{\partial t} \, dV \qquad (7.23)$$

assuming that c and ρ are constants. Transforming the left-hand side by the divergence theorem, we obtain

$$\int\int\int_V \nabla \cdot (\kappa \nabla u) \, dV = \int\int\int_V c\rho \frac{\partial u}{\partial t} \, dV$$

or

$$\int\int\int_V \left\{ \nabla \cdot (\kappa \nabla u) - c\rho \frac{\partial u}{\partial t} \right\} dV = 0. \qquad (7.24)$$

Now, making the further assumption that κ is constant and because V is arbitrary, we finally obtain

$$\nabla \cdot \nabla u = \nabla^2 u = \frac{c\rho}{\kappa} \frac{\partial u}{\partial t}, \qquad (7.25)$$

which is the basic differential equation of heat conduction.

7.5 Poisson's Equation

In this section, we consider a problem dealing with a gravitational field. Let $\mathbf{F}$ be the field intensity at a general point $P(x, y, z)$, and let $\Delta \mathbf{F}$ be the contribution to $\mathbf{F}$ due to the infinitesimal mass Δm_i in an infinitesimal volume $\Delta V_1 = \Delta x_1 \Delta y_1 \Delta z_1$ enclosing the point $P_1(x_1, y_1, z_1)$. According to Newton's law of universal gravitation, $\Delta \mathbf{F}$ is a vector whose magnitude is

$$\Delta \mathbf{F} = k \frac{1 \cdot \Delta m_1}{r^2} \qquad (7.26)$$

where $r^2 = (x - x_1)^2 + (y - y_1)^2 + (z - z_1)^2$ and whose direction is opposite to that of the vector

$$\mathbf{R} = (x - x_1)\mathbf{i} + (y - y_1)\mathbf{j} + (z - z_1)\mathbf{k}$$

extending from P_1 to P (see Figure 7.9).

If we choose $k = -1$ in equation (7.26), then the field intensity at P due to the infinitesimal mass Δm_1 at P_1 can be written in vector form as

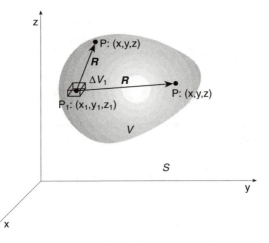

FIGURE 7.9
Potential at point P due to material in element ΔV_1

$$\Delta \mathbf{F} = -\frac{\Delta m_1}{r^2}\frac{\mathbf{R}}{r} = -\rho(x,y,z)\Delta V_1 \frac{\mathbf{R}}{r^3} \qquad (7.27)$$

where $\rho(x,y,z)$ is the density of the material at point P_1.

Now let S be an arbitrary closed regular surface bounding a volume V, and let I denote the integral over S of the normal component of the force due to all attracting material in the field. We know that $\mathbf{F} = \nabla\phi$ where $\phi(x,y,z)$ is the potential function of the field. Therefore, we have

$$I = \int\int_S \mathbf{n}\cdot\mathbf{F}\ dS = \int\int_S \mathbf{n}\cdot\nabla\phi\,dS. \qquad (7.28)$$

However, I can also be computed by first determining the part of it, ΔI, due to the material in an arbitrary volume element ΔV_1 and then taking into account all such infinitesimal contributions by integrating over the entire field. Thus, using (7.27) and (7.28), we obtain

$$\Delta I = \int\int_S \mathbf{n}\cdot\Delta\mathbf{F}\,dS$$
$$= -\int\int_S [\rho(x_1,y_1,z_1)\Delta V_1]\,\mathbf{n}\cdot\frac{\mathbf{R}}{r^3}dS$$
$$= -\rho(x_1,y_1,z_1)\Delta V_1 \int\int_S \mathbf{n}\cdot\frac{\mathbf{R}}{r^3}dS \qquad (7.29)$$

because x_1, y_1 and z_1 are constants with respect to the x, y, z integration over S.

The last integral can, of course, be evaluated by Gauss' Theorem (Stokes' Theorem). Specifically, if the origin of $\mathbf{R}$, namely the point $P_1(x_1, y_1, z_1)$ is within S, the value of the integral is 4π; otherwise, the value of the integral is zero. Hence,

$$\Delta I = \begin{cases} -4\pi\rho(x_1, y_1, z_1)\Delta V_1 & \Delta V_1 \text{ within } S \\ 0 & \Delta V_1 \text{ outside } S \end{cases}$$

and integrating only over the volume V bounded by S, we obtain

$$I = \int dI = -4\pi \int \int \int_V \rho(x_1, y_1, z_1) \, dV_1.$$

Because x_1, y_1, and z_1 are dummy variables, we can write this as

$$I = -4\pi \int \int \int_V \rho(x, y, z) \, dV. \tag{7.30}$$

Equating the two expressions in (7.28) and (7.30), we obtain the following

$$\int \int_S \mathbf{n} \cdot \nabla\phi \, dS = -4\pi \int \int_V \int \rho(x, y, z) \, dV.$$

We can now apply the Divergence Theorem to the integral on the left to get

$$\int \int_V \int \nabla \cdot (\nabla\phi) \, dV = -4\pi \int \int_V \int \rho(x, y, z) \, dV$$

or

$$\int \int_V \int [\nabla \cdot (\nabla\phi) + 4\pi\rho(x, y, z)] \, dV = 0.$$

Because this holds for an arbitrary volume V, it follows that the integrand must vanish identically. This leads to the following equation

$$\nabla^2\phi = -4\pi\rho(x, y, z). \tag{7.31}$$

This equations is know as Poisson's equation.

REMARK 7.3   We have thus shown that in regions occupied by matter, the gravitational potential satisfies Poisson's equation. In empty space $\rho(x, y, z) = 0$, and thus in empty space the gravitational potential satisfies Laplace's equation

$$\nabla^2\phi = 0.$$

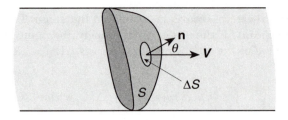

FIGURE 7.10
Charge through a volume element ΔS

7.6   Vectors in Electromagnetic Theory

The theory of electromagnetic fields and waves is a subject where the basic experimental laws concerning electric and magnetic fields can be fully described by five equations. These five equations, in concise form, utilize the divergence and curl concepts of vector analysis. They have equivalent integral forms involving line, surface and volume integrals as discussed in the previous chapters.

It is also possible, using vector identities, to manipulate the basic equations so as to mathematically predict phenomena, such as electromagnetic waves, which are of major importance in, for example, communications engineering. Because electromagnetic field theory is also useful at the "heavy" end of electrical engineering, in connection with electrical machines, are of interest to the physicist, dealing with applications of high-intensity magnetic fields, and to the biologist concerned with modelling electrical behavior of the heart and brain. It is clear that it is an area of far too many wide-ranging applications to discuss.

It is possible to study electromagnetic phenomena without using vector analysis, but the mathematical equations quickly become long and involved and the underlying physics is easily lost. With the aid of vector analysis, we can readily represent the basic equations in a compact and concise manner and, hopefully, understand their physical significance.

7.7   The Continuity Equation Revisited

Electric current is the flow of electric charge. The electric current across a surface S is defined as the rate at which electric charge flows across S.

If q denotes the electric charge density, i.e., the amount of electric charge per unit volume, and $\mathbf{V}$ is the velocity of the electric charge, then the vector field

$$\mathbf{J} = Q\mathbf{V}$$

is called the **current density**. The current I crossing a surface element ΔS in an electric conductor is given by

$$I = Q|\mathbf{V}|\cos\theta\Delta S = Q\mathbf{V}\cdot\mathbf{n}\Delta S = \mathbf{J}\cdot\mathbf{n}\Delta S$$

because this is the amount of charge that will cross ΔS every second (see Figure 7.10).

For a nonelementary surface S, the current will then be given by

$$I = \int\int_S \mathbf{J}\cdot\mathbf{n}\,dS = \int\int_S \mathbf{J}\cdot d\mathbf{S}, \tag{7.32}$$

which is recognized as a flux integral. Clearly, the sign of I depends on the sense of the unit normal $\mathbf{n}$. If S is a closed surface, $\mathbf{n}$ is chosen to be the outwardly directed unit normal. As in the corresponding calculation for fluid flow, the continuity equation for current flow is derived using a simple physical principle, in this case the principle of conservation of electric charge. The total charge q in a volume V is given by the volume integral of the charge density,

$$q = \int\int\int_V Q\,dV$$

so that the rate at which charge decreases in this volume is

$$-\frac{\partial q}{\partial t} = -\frac{\partial}{\partial t}\int\int\int_V Q\,dV = -\int\int\int_V \frac{\partial q}{\partial t}\,dV.$$

But, the total electric charge in V can only change if charge is flowing across the bounding surface S. Hence, we must equate the rate of flow of charge across S as given by (7.32) with the rate of decrease of charge in V. In other words,

$$\int\int_S \mathbf{J}\cdot\mathbf{n}\,dS = -\int\int\int_V \frac{\partial Q}{\partial t}\,dV.$$

Transforming the left-hand side with the aid of the divergence theorem, we obtain

$$\int\int\int_V \nabla\cdot\,dV = -\int\int\int_V \frac{\partial Q}{\partial t}\,dV$$

or

$$\int\int\int_V \left(\nabla \cdot \mathbf{J} + \frac{\partial Q}{\partial t}\right) dV = 0. \qquad (7.33)$$

However, since we require that (7.33) hold for an arbitrary region V, the integrand must be zero at each point, i.e.,

$$\nabla \cdot \mathbf{J} = -\frac{\partial Q}{\partial t}. \qquad (7.34)$$

This is the continuity equation for current flow. As one would expect, it is identical in form with the continuity equation for fluid flow. This is the first of the five equations.

7.8   Maxwell's Equations for Electromagnetic Fields

The remaining four equations of electromagnetism will be discussed in this section. These equations involve the following basic field vectors:

- the electric field $\mathbf{E}$

- the electric flux density $\mathbf{D}$

- the magnetic field $\mathbf{H}$

- the magnetic flux density $\mathbf{B}$

- the current density $\mathbf{J}$

and the scalars:

$$\epsilon = \text{permittivity}$$
$$\mu = \text{permeability}$$
$$\sigma = \text{conductivity}$$
$$Q = \text{charge density}$$
$$q = \int\int\int_V Q\, dV = \text{ total charge within volume } V$$
$$\phi = \int\int_S \mathbf{n} \cdot \mathbf{B}\, dS = \text{ total magnetic flux passing through surface } S$$
$$i = \int\int_S \mathbf{n} \cdot \mathbf{J}\, dS = \text{ total current flowing through surface } S.$$

The field vectors **D** and **H** are related to **E** and **B**, respectively, by functional relationships which are characteristic of the medium. The simplest situation is where the relationships are

$$\mathbf{D} = \epsilon\mathbf{E} \tag{7.35}$$

and

$$\mathbf{H} = \frac{1}{\mu}\mathbf{B} \tag{7.36}$$

where ϵ and μ are the scalars previously defined. The quantities just defined (field vectors and scalars) are connected by a number of equations expressing relations discovered experimentally in the early nineteenth century chiefly by Michael Faraday (1791–1867). In particular, we have

1. **Faraday's Law,**

$$\int_C \mathbf{E} \cdot d\mathbf{R} = -\frac{\partial\phi}{\partial t} \tag{7.37}$$

    which asserts that the integral of the tangential component of the electric intensity vector around any closed curve C is equal but opposite in sign to the rate of change of the magnetic flux passing through any surface C.

2. **Ampere's Law,**

$$\int_C \mathbf{H} \cdot d\mathbf{R} = i \tag{7.38}$$

    which asserts that the integral of the tangential component of the magnetic intensity vector around any closed curve C is equal to the current flowing through any surface spanning C.

3. **Gauss' Law for electric fields,**

$$\int\int_S \mathbf{N} \cdot \mathbf{D}\, dS = q \tag{7.39}$$

    which asserts that the integral of the normal component of the electric flux density over any closed surface S is equal to the total electric charge enclosed by S.

4. **Gauss' Law for magnetic fields,**

$$\int\int_S \mathbf{N} \cdot \mathbf{B}\, dS = 0 \tag{7.40}$$

which asserts that the total magnetic flux passing through any closed surface S is zero. We can now state, in general form, Maxwell's equations:

(a) Divergence equations:

$$\nabla \cdot \mathbf{D} = Q \tag{7.41}$$
$$\nabla \cdot \mathbf{B} = 0. \tag{7.42}$$

(b) Curl equations:

$$\nabla \times \mathbf{E} = -\frac{\partial \mathbf{B}}{\partial t} \tag{7.43}$$
$$\nabla \times \mathbf{H} = \mathbf{J} + \frac{\partial \mathbf{D}}{\partial t}. \tag{7.44}$$

To show how these are obtained, we begin by applying the divergence theorem to Gauss' first law (7.39) to get:

$$\int \int \int_V \nabla \cdot \mathbf{D} \, dV = q = \int \int \int_V Q \, dV$$

from which, since V is arbitrary, it follows that

$$\nabla \cdot \mathbf{D} = Q. \tag{7.45}$$

In the same way, by applying the divergence theorem to Gauss' second law (7.40), we find that

$$\int \int \int_V \nabla \cdot \mathbf{B} \, dV = 0$$

and because V is arbitrary,

$$\nabla \cdot \mathbf{B} = 0.$$

Next, if we apply Stokes' theorem to Faraday's Law (7.37), we have

$$\int \int_S \mathbf{N} \cdot \nabla \times \mathbf{E} \, dS = -\frac{\partial \phi}{\partial t},$$

and substituting for ϕ from its definition in terms of $\mathbf{B}$,

$$\int \int_S \mathbf{N} \cdot \nabla \mathbf{E} \, dS = -\frac{\partial}{\partial t} \left(\int \int_S \mathbf{N} \cdot \mathbf{B} \, dS \right) = -\int \int_S \mathbf{N} \cdot \frac{\partial \mathbf{B}}{\partial t} \, dS.$$

Because S is an arbitrary surface spanning the arbitrary closed curve C, the last equation can hold if

$$\nabla \times \mathbf{E} = -\frac{\partial \mathbf{B}}{\partial t}.$$

Similarly, if we apply Stokes' theorem to Ampere's Law (7.38), we obtain

$$\int\int_S \mathbf{N} \cdot \nabla \times \mathbf{H} \, dS = i = \int\int_S \mathbf{N} \cdot \mathbf{J} \, dS$$

and, again, because S is an arbitrary open surface, we conclude that the vectors being integrated over S must be identical, i.e.,

$$\nabla \times \mathbf{H} = \mathbf{J}. \tag{7.46}$$

As Maxwell was the first to realize, the current density $\mathbf{J}$ consists of two parts, namely, a conduction current density,

$$\mathbf{J}_c = \sigma \mathbf{E},$$

due to the flow of electric charges, and a displacement(electric flux) current density,

$$\mathbf{J}_d = \frac{\partial \mathbf{D}}{\partial t} = \epsilon \frac{\partial \mathbf{E}}{\partial t},$$

due to the time variation of electric field. Thus,

$$\mathbf{J} = \sigma \mathbf{E} + \epsilon \frac{\partial \mathbf{E}}{\partial t}$$

and (7.46) becomes

$$\nabla \times \mathbf{H} = \sigma \mathbf{E} + \epsilon \frac{\partial \mathbf{E}}{\partial t} \tag{7.47}$$

or

$$\nabla \times \mathbf{H} = \mathbf{J}_c + \frac{\partial \mathbf{D}}{\partial t} \tag{7.48}$$

where $\mathbf{J}_c$ is the electric current density. If we take the curl of (7.43) we obtain

$$\nabla \times (\nabla \times \mathbf{E}) = -\nabla \times \frac{\partial \mathbf{B}}{\partial t} = -\frac{\partial}{\partial t}(\nabla \times \mathbf{B})$$

$$= -\mu \frac{\partial}{\partial t}(\nabla \times \mathbf{H}).$$

If the term $\nabla \times (\nabla \times \mathbf{E})$ is expanded, the last equation becomes

$$\nabla(\nabla \cdot \mathbf{E}) - \nabla^2 \mathbf{E} = -\mu \frac{\partial}{\partial t} (\nabla \times \mathbf{H})$$

and, substituting for $\nabla \times \mathbf{H}$ from (7.47),

$$\nabla(\nabla \cdot \mathbf{E}) - \nabla^2 \mathbf{E} = -\mu \frac{\partial}{\partial t} \left(\sigma \mathbf{E} + \epsilon \frac{\partial \mathbf{E}}{\partial t} \right). \qquad (7.49)$$

If the space charge density Q is zero, as it is to a high degree of approximation in both dielectrics and good conductors, then from (7.45) and the relation $\mathbf{D} = \epsilon \mathbf{E}$ we see that

$$\nabla \cdot \mathbf{E} = 0.$$

Therefore, equation (7.49) reduces to

$$\nabla^2 \mathbf{E} = \mu\epsilon \frac{\partial^2 \mathbf{E}}{\partial t^2} + \mu\sigma \frac{\partial \mathbf{E}}{\partial t},$$

which is **Maxwell's equation for the electric intensity vector E**. In a similar manner, if we take the curl of (7.47), we obtain

$$\nabla \times (\nabla \times \mathbf{H}) = \nabla \times \left(\sigma \mathbf{E} + \epsilon \frac{\partial \mathbf{E}}{\partial t} \right)$$

and, expanding the left-hand side,

$$\nabla(\nabla \cdot \mathbf{H}) - \nabla^2 \mathbf{H} = \sigma \nabla \times \mathbf{E} + \epsilon \nabla \times \frac{\partial \mathbf{E}}{\partial t}$$

$$= \sigma \nabla \mathbf{E} + \epsilon \frac{\partial}{\partial t} (\nabla \times \mathbf{E}).$$

Now substituting for $\nabla \times \mathbf{E}$ from (7.43)

$$\nabla(\nabla \cdot \mathbf{H}) - \nabla^2 \mathbf{H} = \sigma \left(-\frac{\partial \mathbf{B}}{\partial t} \right) + \epsilon \left(-\frac{\partial^2 \mathbf{B}}{\partial t^2} \right).$$

But, $\mathbf{B} = \mu \mathbf{H}$, by definition. Hence (7.42) implies that $\nabla \cdot \mathbf{H} = 0$, and therefore the last equation reduces to

$$\nabla^2 \mathbf{H} = \mu\epsilon \frac{\partial^2 \mathbf{H}}{\partial t^2} + \mu\sigma \frac{\partial \mathbf{H}}{\partial t},$$

which is **Maxwell's equation for the magnetic intensity vector H**.

For a perfect dielectric, $\sigma = 0$. Hence, in this case, Maxwell's equations reduce to the one-dimensional wave equations

$$\nabla^2 \mathbf{E} = \mu\epsilon \frac{\partial^2 \mathbf{E}}{\partial t^2} \text{ and } \nabla^2 \mathbf{H} = \mu\epsilon \frac{\partial^2 \mathbf{H}}{\partial t^2}.$$

On the other hand, in a good conductor the terms arising from the displacement current, i.e., the terms containing the second time derivatives, are negligible, and Maxwell's equations reduce to

$$\nabla^2 \mathbf{E} = \mu\sigma \frac{\partial \mathbf{E}}{\partial t} \text{ and } \nabla^2 \mathbf{H} = \mu\sigma \frac{\partial \mathbf{H}}{\partial t},$$

which are examples of the three-dimensional heat equation.

Both the curl equations (7.43), (7.44) relate an electric vector ($\mathbf{E}$ or $\mathbf{D}$) with a magnetic vector ($\mathbf{B}$ or $\mathbf{H}$), justifying the use of the term "electromagnetic". However, for fields that are classed as static or time-independent, the time-derivative terms are zero and the curl equations become

$$\nabla \times \mathbf{E} = 0 \tag{7.50}$$

$$\nabla \times \mathbf{H} = \mathbf{J}_c. \tag{7.51}$$

That is, the electric and magnetic vectors are separate or **uncoupled** in the static situation. The theory of static fields involves the study of equations (7.50), (7.51) and (7.41), (7.42), which are the same for static and time varying fields.

The interpretation of (7.41) for electric fields is that positive charge is a **source** of the vector $\mathbf{D}$ and negative charge is a **sink** ($\nabla \cdot \mathbf{D} > 0$ when $q > 0$ and $\nabla \cdot \mathbf{D} < 0$ when $q < 0$). At points not occupied by electric charge, $\nabla \cdot \mathbf{D}$ is zero. Equation (7.50), for static fields, tells us that an electrostatic field $\mathbf{E}$ is an example of a conservative vector field so that

$$\int_C \mathbf{E} \cdot d\mathbf{R} = 0$$

for any closed path C. Physically, because $\mathbf{E}$ is the force on unit charge, the zero value for this line integral tells us that the work done in moving a unit charge around a closed path is always zero. It follows that we can neither gain nor lose energy by such a process. Also, because $\mathbf{E}$ is conservative, we can represent $\mathbf{E}$ as the gradient of a scalar potential ϕ. Thus, we can write

$$\mathbf{E} = -\nabla\phi \text{ (static fields)}. \tag{7.52}$$

The advantage of using ϕ in electrostatic field problems is that it is usually easier to calculate a scalar quantity than a vector. To actually find ϕ (and then $\mathbf{E}$, using (7.52)), we must determine the equation satisfied by ϕ. This is readily done using (7.41) and, assuming that the medium is such that $\mathbf{D} = \epsilon\mathbf{E}$ holds:

$$\nabla \cdot \mathbf{D} = \nabla \cdot (\epsilon \mathbf{E}) = \epsilon \nabla \cdot \mathbf{E} = -\epsilon \nabla \cdot \nabla \phi = Q$$

or

$$\nabla^2 \phi = -\frac{Q}{\epsilon}. \tag{7.53}$$

Equation (7.53), which is called **Poisson's equation**, is the relation between the potential at a point and the electric charge density at that point. In particular, at points where no charge is present $(Q = 0)$, the potential satisfies

$$\nabla^2 \phi = 0 \tag{7.54}$$

which is **Laplace's equation**. Practical electrostatic problems often involve solving Laplace's equation in a region with known boundary conditions, for example calculating the potential and field between the conductors in a capacitor. The equation satisfied by the electric field $\mathbf{E}$ at points where $Q = 0$ is, of course,

$$\nabla \cdot \mathbf{E} = 0. \tag{7.55}$$

Example 7.1

Given the vector field

$$\mathbf{F} = \frac{K}{r^n} \mathbf{r} \ (r \neq 0)$$

(in spherical coordinates), deduce the value of n that makes the field solenoidal. Show that $\mathbf{F}$ is conservative for any value of n. Give an interpretation for $\mathbf{F}$ in an electrostatic context.   ▯

SOLUTION   Using spherical coordinates, the divergence is

$$\nabla \cdot \mathbf{F} = \frac{1}{r^2} \frac{\partial}{\partial r} \left(\frac{Kr^2}{r^n} \right) = \frac{K}{r^2} (2 - n) r^{1-n} = K \left(\frac{2 - n}{r^{n+1}} \right),$$

which is solenoidal $(\nabla \cdot \mathbf{F} = 0)$ if $n = 2$. The curl in spherical coordinates is given by

$$\nabla \times \mathbf{F} = \frac{1}{r^2 \sin \theta} \begin{vmatrix} \mathbf{r} & r\boldsymbol{\theta} & r \sin \theta \boldsymbol{\phi} \\ \frac{\partial}{\partial r} & \frac{\partial}{\partial \theta} & \frac{\partial}{\partial \phi} \\ \frac{K}{r^n} & 0 & 0 \end{vmatrix}$$

$$= \mathbf{0} \quad \text{(all components zero)}.$$

Because $\mathbf{F}$ has zero divergence when $n = 2$ and also zero curl, it follows from (7.50) and (7.55) that

$$\mathbf{F} = \frac{K}{r^2}\mathbf{r}$$

is a possible electrostatic field at any point $r \neq 0$. In fact, $\mathbf{F}$ represents the field $\mathbf{E}$ due to a point charge at $r = 0$, the value of the constant K depending on the system of units being employed. ∎

Turning now to equations (7.42) and (7.51) for the magnetostatic field, the first of these,

$$\nabla \cdot \mathbf{B} = 0,$$

tells us that a magnetic field is always solenoidal, so that magnetic field lines neither start nor end. This implies that isolated magnetic poles do not exist (i.e., there is no magnetic equivalent of an electric charge). The equation

$$\nabla \times \mathbf{H} = \mathbf{J}_c$$

links the magnetic field $\mathbf{H}$ and the electric current density $\mathbf{J}_c$ at a point. Because $\nabla \times \mathbf{H}$ is only zero at points where $\mathbf{J} = 0$, that is, outside current-carrying conductors, it is clear that $\mathbf{H}$, unlike $\mathbf{E}$, is in general a non-conservative vector field. Therefore, we cannot normally represent $\mathbf{H}$ as the gradient of a scalar potential, although a magnetic scalar potential function is occasionally used to solve magnetic field problems in regions where $\mathbf{J}_c$ is zero everywhere.

Example 7.2
Deduce the corresponding integral forms of the Maxwell equations (7.41), (7.42) and (7.50), (7.51) for static fields, and interpret the results. ⧠

SOLUTION   For the divergence equations (7.41) and (7.42), we integrate both sides of each equation over an arbitrary volume V bounded by a surface S and use the divergence theorem. In the electrostatic case we obtain

$$\int \int \int_V \nabla \cdot \mathbf{D} \, dV = \int \int \int_V Q \, dV$$

or

$$\int \int_S \mathbf{D} \cdot \mathbf{n} \, dS = q \tag{7.56}$$

where $q = \int \int \int_V Q \, dV$ is the total charge in the volume V. Equation (7.56) tells us that the electric flux over a closed surface equals the charge inside that surface. This is known as **Gauss' Law**, and it can be used for finding

D (or **E**) in the field problems where a high degree of symmetry is present. Similarly, equation (7.42) gives

$$\int \int \int_V \nabla \cdot \mathbf{B} \, dV = 0 \text{ for any volume } V,$$

which transforms to

$$\int \int_S \mathbf{B} \cdot \mathbf{n} \, dS = 0. \tag{7.57}$$

This is Gauss' law in magnetostatics. It tells us that the magnetic flux over any closed surface is zero, confirming the nonexistence of sources or sinks for magnetic fields. In fact, (7.57) holds for any magnetic field, static or time-varying, because the result $\nabla \cdot \mathbf{B} = 0$ is valid in both cases.

For the curl equations (7.50) and (7.51), we find the flux of both sides over an arbitrary surface S bounded by a closed curve C. The magnetic field equation (7.51) gives

$$\int \int_S \nabla \times \mathbf{H} \cdot \mathbf{n} \, dS = \int \int_S \mathbf{J}_c \cdot \mathbf{n} \, dS$$

or, using Stokes' theorem to transform the left-hand side,

$$\oint_C \mathbf{H} \cdot d\mathbf{R} = I \tag{7.58}$$

where we have also used (7.32) to identify the right-hand side as the electric current enclosed by the path C. Equation (7.58) is known as **Ampere's law in magnetostatics**; it can be used to find the magnitude of a magnetic field in a problem where a high degree of symmetry is present. Similarly, for the electrostatic field, equation (7.50) transforms (by Stokes' theorem) into

$$\oint_C \mathbf{E} \cdot d\mathbf{R} = 0,$$

which merely confirms that the electrostatic field is conservative. ∎

7.9   Waves Solutions of Maxwell's Equations

One consequence of the Maxwell equations for time-varying electromagnetic fields is the prediction that variations in the electric and magnetic fields **E** and **H** propagate as waves. The formal analysis leading to this prediction is readily carried out using the tools of vector analysis. The experimental verification of the predictions by Hertz was powerful evidence for the validity of the Maxwell equations. The simplest model to consider is where the Maxwell equations (7.41)–(7.44) take on the forms

$$\nabla \cdot \mathbf{D} = 0$$
$$\nabla \cdot \mathbf{B} = 0$$
$$\nabla \times \mathbf{E} = -\frac{\partial \mathbf{B}}{\partial t}$$
$$\nabla \times \mathbf{H} = \frac{\partial \mathbf{D}}{\partial t}$$

or, eliminating $\mathbf{D}$ and $\mathbf{B}$,

$$\nabla \cdot \mathbf{E} = 0 \tag{7.59}$$
$$\nabla \cdot \mathbf{H} = 0 \tag{7.60}$$
$$\nabla \times \mathbf{E} = -\mu \frac{\partial \mathbf{H}}{\partial t} \tag{7.61}$$
$$\nabla \times \mathbf{H} = \epsilon \frac{\partial \mathbf{E}}{\partial t}. \tag{7.62}$$

From these equations it is evident that we are considering a linear homogeneous medium with no electric charges or electric currents at any point. The symmetry of the two divergence equations (7.59), (7.60) and the two curl equations (7.61), (7.62) is readily seen.

To obtain predictions of wave behavior, we take the curl of both sides of equation (7.61) and use the identity for $\nabla \times \nabla \times \mathbf{A}$. We obtain

$$\nabla(\nabla \cdot \mathbf{E}) - \nabla^2 \mathbf{E} = -\mu \frac{\partial}{\partial t} (\nabla \times \mathbf{H})$$

or

$$\nabla^2 \mathbf{E} = \mu\epsilon \frac{\partial^2 \mathbf{E}}{\partial t^2} \tag{7.63}$$

using (7.59) and (7.62). A similar procedure using the curl of equation (7.62) yields

$$\nabla^2 \mathbf{H} = \mu\epsilon \frac{\partial^2 \mathbf{H}}{\partial t^2}. \tag{7.64}$$

An alternate and more compact form of the wave equation is

$$\frac{\partial^2 \mathbf{E}}{\partial t^2} = c^2 \nabla^2 \mathbf{E} \tag{7.65}$$

where $c^2 = 1/\mu_0\varepsilon_0$. Equations (7.63) and (7.64) are both vector equations and are thus each equivalent to three scalar equations. In Cartesian form, these equations are

$$\nabla^2 E_x = \mu\epsilon\frac{\partial E_x}{\partial t^2} \text{ and } \nabla^2 H_x = \mu\epsilon\frac{\partial H_x}{\partial t^2} \qquad (7.66)$$

with similar relationships for the y- and z-components. Equations (7.66) and (7.65) are called **wave equations** because their solutions are waves travelling at speed c. The constant c is the speed of light and the waves are known as electromagnetic waves.

Many different solutions, known as **plane waves**, are considered here. In this case, the magnitudes of the vectors $\mathbf{E}$ and $\mathbf{H}$ are assumed to be constant at all points on a plane perpendicular to a given direction, which we will take to be the z-axis of a Cartesian coordinate system. Mathematically, this implies that all partial derivatives with respect to x and y may be set to zero.

Example 7.3
Suppose that $\mathbf{E} = (f(x,t), 0, 0)$ in a Cartesian coordinate system so that the wave equation becomes

$$\frac{\partial^2 f}{\partial t^2} = c^2\frac{\partial^2 f}{\partial x^2}. \qquad (7.67)$$

Show that $f(x,t) = \sin k(x - ct)$ is a solution for any value of the constant k and interpret this solution physically.   ⬚

SOLUTION   For the function $f(x,t) = \sin k(x - ct)$,

$$\frac{\partial^2 f}{\partial t^2} = -(kc)^2 \sin k(x - ct) \text{ and } \frac{\partial^2 f}{\partial x^2} = -k^2 \sin k(x - ct),$$

so $f(x,t) = \sin k(x - ct)$ obeys (7.67). Physically, this solution corresponds to a sine wave travelling to the right at speed c. At $t = 0$, $f(x,0) = \sin kx$, which is a sine wave that has $f = 0$ at $x = 0$. At a later time t, the point at which $f = 0$ has moved to the position $x = ct$, so the wave has moved to the right a distance ct (Figure 7.11). The speed of the wave is therefore c. Similarly, it can be shown that the function $g(x,t) = \sin k(x + ct)$ is a solution of (7.67), representing a sine wave travelling to the left at speed c. In Figure 7.11, the solid line represents the solution at $t = 0$, while the dashed line represents the solution at a later time.   ∎

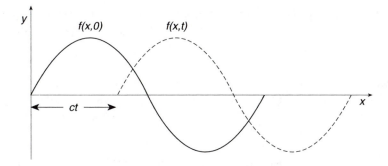

FIGURE 7.11
The solution $f(x,t) = \sin k(x - ct)$ of the wave equation

Example 7.4
Show that $\mathbf{E} = \mathbf{E}_0 f(\mathbf{k} \cdot \mathbf{x} - \omega t)$ is a solution to the wave equation (7.65), where $\mathbf{E}_0$ and $\mathbf{k}$ are constant vectors, f is any function and $\omega = \pm c|\mathbf{k}|$. ▯

SOLUTION   Let $u = \mathbf{k} \cdot \mathbf{x} - \omega t$ so that the solution under consideration is $\mathbf{E} = \mathbf{E}_0 f(u)$. Then

$$\frac{\partial \mathbf{E}}{\partial t} = \mathbf{E}_0 \frac{df}{du} \frac{\partial u}{\partial t} = -\omega \mathbf{E}_0 \frac{df}{du},$$

and similarly

$$\frac{\partial^2 \mathbf{E}}{\partial t^2} = -\omega \mathbf{E}_0 \frac{d^2 f}{du^2} \frac{\partial u}{\partial t} = \omega^2 \mathbf{E}_0 \frac{d^2 f}{du^2}.$$

To find $\nabla^2 \mathbf{E}$, note that because $\mathbf{E}_0$ is constant,

$$\nabla^2 (\mathbf{E}_0 f(u)) = \mathbf{E}_0 \nabla^2 f(u) = \mathbf{E}_0 \nabla \cdot (\nabla f(u)).$$

The gradient of $f(u)$ is

$$\nabla f(u) = \frac{df}{du} \nabla (\mathbf{k} \cdot \mathbf{x} - \omega t) = \frac{df}{du} \mathbf{k},$$

and taking the divergence of this gives

$$\nabla^2 f(u) = \nabla \left(\frac{df}{du} \right) \cdot \mathbf{k} = \frac{d^2 f}{du^2} \nabla u \cdot \mathbf{k} = \frac{d^2 f}{du^2} |\mathbf{k}|^2.$$

Therefore, the wave equation is obeyed provided that

$$\omega^2 \mathbf{E}_0 \frac{d^2 f}{du^2} = c^2 \mathbf{E}_0 \frac{d^2 f}{du^2} |\mathbf{k}|^2 \quad \text{which gives} \quad \omega^2 = c^2 |\mathbf{k}|^2.$$

So the function f is arbitrary and the only condition is that the frequency of the wave ω must be related to the speed c and the constant vector $\mathbf{k}$ (that is known as the wave vector) by $\omega = \pm c|\mathbf{k}|$. ■

The Maxwell equation $\nabla \cdot \mathbf{E} = 0$ reduces, for this case, to $\partial E_z/\partial z = 0$, which tells us that E_z does not vary with z. Hence, E_z is constant in space and we can, without loss of generality, take it to be zero. Therefore, we need only consider an electric field vector of the form

$$\mathbf{E}(z,t) = E_x(z,t)\mathbf{i} + E_y(z,t)\mathbf{j}$$

and, by similar reasoning, because $\nabla \cdot \mathbf{H} = 0$, we assume a magnetic field vector

$$\mathbf{H}(z,t) = H_x(z,t)\mathbf{i} + H_y(z,t)\mathbf{j}.$$

The equations satisfied by the components of $\mathbf{E}$ and $\mathbf{H}$ are, using (7.66)

$$\frac{\partial^2 E_x}{\partial z^2} = \mu\epsilon\frac{\partial^2 E_x}{\partial t^2} \tag{7.68}$$

$$\frac{\partial^2 H_x}{\partial z^2} = \mu\epsilon\frac{\partial^2 H_x}{\partial t^2} \tag{7.69}$$

and similarly for the y-components. These equations are all examples of the one-dimensional wave equation. It is easy to show, by substitution, that

$$E_x = E_{x0}f(z - vt) \tag{7.70}$$

(where $v = 1/\sqrt{(\mu\epsilon)}$) is one possible solution of equation (7.68), where f is any well-behaved (differentiable) function of the variable $(z - vt)$ and E_{x0} is a constant. Physically, such a solution represents a wave travelling without change of shape in the positive z-direction with a speed of v (see Figure 7.12).

Similarly, one solution for the y-component of $\mathbf{E}$ is

$$E_y = E_{y0}f(z - vt)$$

we see that the $\mathbf{E}$ vector in a plane electromagnetic wave is normal or transverse to the direction in which the wave is propagating. We now calculate the corresponding components of $\mathbf{H}$ using the Maxwell equation (7.61) and the Cartesian form of the curl:

$$\nabla \times \mathbf{E} = -\frac{\partial E_y}{\partial z}\mathbf{i} + \frac{\partial E_x}{\partial z}\mathbf{j}.$$

Hence, since

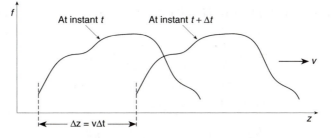

FIGURE 7.12
Wave travelling with velocity v

$$\frac{\partial \mathbf{H}}{\partial t} = \frac{\partial H_x}{\partial t}\mathbf{i} + \frac{\partial H_y}{\partial t}\mathbf{j}$$

we obtain, on comparing components

$$\frac{\partial E_y}{\partial z} = \mu \frac{\partial H_x}{\partial t} \tag{7.71}$$

$$\frac{\partial E_x}{\partial z} = -\mu \frac{\partial H_y}{\partial t}. \tag{7.72}$$

But from (7.70),

$$\frac{\partial E_x}{\partial z} = \frac{\partial E_x}{\partial (z - vt)}\frac{\partial (z - vt)}{\partial z} = \frac{\partial E_x}{\partial (z - vt)}$$

and

$$\frac{\partial E_x}{\partial t} = \frac{\partial E_x}{\partial (z - vt)}\frac{\partial (z - vt)}{\partial t} = -v\frac{\partial E_x}{\partial (z - vt)}$$

from which

$$\frac{\partial E_x}{\partial z} = -\frac{1}{v}\frac{\partial E_x}{\partial t}.$$

Hence (7.72) becomes

$$\frac{\partial H_y}{\partial t} = \frac{1}{v\mu}\frac{\partial E_x}{\partial t} = \sqrt{\left(\frac{\epsilon}{\mu}\right)}\frac{\partial E_x}{\partial t}$$

and integrating

$$H_y = \sqrt{\left(\frac{\epsilon}{\mu}\right)}E_x = \sqrt{\left(\frac{\epsilon}{\mu}\right)}E_{x0}f(z - vt).$$

Similar reasoning applied to (7.71) leads to

$$H_x = -\sqrt{\left(\frac{\epsilon}{\mu}\right)} E_y = -\sqrt{\left(\frac{\epsilon}{\mu}\right)} E_{y0} f(z - vt).$$

We see that variations of **H**, line those of **E**, propagate as transverse travel-ling waves with speed $v = 1/\sqrt{(\epsilon\mu)}$. The ratio

$$\frac{|\mathbf{E}|}{|\mathbf{H}|} = \frac{\sqrt{(E_x^{\,2} + E_y^{\,2})}}{\sqrt{(H_x^{\,2} + H_y^{\,2})}} = \frac{\sqrt{(E_x^2 + E_y^2)}}{\sqrt{\left(\frac{\epsilon}{\mu}E_x^2 + \frac{\epsilon}{\mu}E_y^2\right)}} = \sqrt{\left(\frac{\mu}{\epsilon}\right)}$$

depends, like the speed, on the electrical constants μ and ϵ. This ratio is referred to as the **characteristic impedance** Z of the medium of propa-gation. For a vacuum, or free space, the numerical values are

$$\epsilon = \epsilon_0 = 8.85 \times 10^{-12} Fm^{-1} \text{ and } \mu = \mu_0 = 4\pi \times 10^{-7} Hm^{-1}$$

from which

$$Z = z_0 = \sqrt{\left(\frac{\mu_0}{\epsilon_0}\right)} \approx 120\pi \text{ ohms and } v = \frac{1}{\sqrt{(\epsilon\mu)}} \approx 3 \times 10^8 ms^{-1}$$

we see that the predicted speed of propagation v the electromagnetic waves in free space is equal to the experimental speed of light. This result was early evidence that light waves are electromagnetic in nature.

Finally, we can easily show that **E** and **H** are mutually perpendicular vectors and that $\mathbf{E} \times \mathbf{H}$ is a vector in the direction of propagation of the electromagnetic wave. The former is shown by noting that

$$\mathbf{E} \cdot \mathbf{H} = E_x H_x + E_y H_y = -E_x \sqrt{\left(\frac{\epsilon}{\mu}\right)} E_y + E_y \sqrt{\left(\frac{\epsilon}{\mu}\right)} E_x = 0$$

while

$$\mathbf{E} \times \mathbf{H} = \begin{vmatrix} \mathbf{i} & \mathbf{j} & \mathbf{k} \\ E_x & E_y & 0 \\ H_x & H_y & 0 \end{vmatrix}$$

$$= (E_x H_y - E_y H_x)\mathbf{k}$$

$$= (E_x^2 + E_y^2)\sqrt{\left(\frac{\epsilon}{\mu}\right)}\mathbf{k}$$

$$= \frac{|\mathbf{E}|^2}{Z}\mathbf{k}$$

so that $\mathbf{E} \times \mathbf{H}$ is indeed directed along the propagation direction (positive z).

Example 7.5

(a) A plane sinusoidal electromagnetic wave in free space has an electric field

$$\mathbf{E} = E_0 \cos(\beta z - \omega t)\mathbf{i}.$$

Deduce the corresponding magnetic field $\mathbf{H}$ and show that the phase velocity of the wave is given by $v = 1/\sqrt{(\mu_0 \epsilon_0)}$.

(b) If $E_0 = 200\pi V m^{-1}$, find the amplitude of the magnetic field and the value of the angular frequency ω if the wavelength is 1.89 m.

$\square$

SOLUTION In this problem we are considering, perhaps, the most important type of plane wave, namely one where the general function f of the earlier theory is sinusoidal, so that instead of $f(z-vt)$ we have $\cos 2\pi(z-vt)/\lambda$, where λ is called the **wavelength**. The name arises because, for a given t,

$$\cos \frac{2\pi}{\lambda}[(z + \lambda) - vt] = \cos \frac{2\pi}{\lambda}(z - vt),$$

that is, the wave repeats after a distance λ. Also if, at a given value of z, t increases by λ/v,

$$\cos \frac{2\pi}{\lambda}\left\{z - v\left(t + \frac{\lambda}{v}\right)\right\} = \cos \left\{\frac{2\pi}{\lambda}(z - vt) - 2\pi\right\} = \cos \frac{2\pi}{\lambda}(z - vt).$$

That is, the period T of the oscillation at a point is λ/v and the frequency f is $1/T = v/\lambda$. The **angular frequency** ω is defined as $2\pi f$. If we also put $2\pi/\lambda = \beta$, we obtain

$$\cos \frac{2\pi}{\lambda}(z - vt) = \cos(\beta z - \omega t) \text{ (or, equivalently, } \cos(\omega t - \beta z)),$$

which is the given form.

(a) We see that, for the given field $\mathbf{E}$,

$$\nabla \cdot \mathbf{E} = \frac{\partial E_x}{\partial x} + \frac{\partial E_y}{\partial y} + \frac{\partial E_z}{\partial z} = \frac{\partial E_x}{\partial x} = 0$$

so that $\mathbf{E}$ does satisfy the Maxwell equation (7.59). To find $\mathbf{H}$ we use (7.61), that is

$$\frac{\partial \mathbf{H}}{\partial t} = -\frac{1}{\mu}\nabla \times \mathbf{E} = -\frac{1}{\mu}\begin{vmatrix} \mathbf{i} & \mathbf{j} & \mathbf{k} \\ \frac{\partial}{\partial x} & \frac{\partial}{\partial y} & \frac{\partial}{\partial z} \\ E_0\cos(\omega t - \beta z) & 0 & 0 \end{vmatrix}$$

$$= -\frac{E_0}{\mu}\beta \sin(\omega t - \beta z)\mathbf{j}.$$

Integrating with respect to t,

$$\mathbf{H} = \frac{\beta E_0}{\omega\mu}\cos(\omega t - \beta z)\mathbf{j}$$

$$= H_0\cos(\omega t - \beta z)\mathbf{j}, \quad \text{say}.$$

Thus, as in the general theory, we have an $\mathbf{H}$ vector with a y-component associated with an $\mathbf{E}$ vector with an x-component, both transverse to the direction of propagation. Clearly,

$$\nabla \cdot \mathbf{H} = \frac{\partial H_y}{\partial y} = 0,$$

i.e., $\mathbf{H}$ satisfies (7.60). Also,

$$\nabla \times \mathbf{H} = \begin{vmatrix} \mathbf{i} & \mathbf{j} & \mathbf{k} \\ \frac{\partial}{\partial x} & \frac{\partial}{\partial y} & \frac{\partial}{\partial z} \\ 0 & H_0\cos(\omega t - \beta z) & 0 \end{vmatrix}$$

$$= \beta H_0 \sin(\omega t - \beta z)\mathbf{i}.$$

But $\partial \mathbf{E}/\partial t = \omega E_0 \sin(\omega t - \beta z)\mathbf{i}$, so that for the final Maxwell equation (7.62) to be satisfied we must have

$$\beta H_0 = \frac{\beta^2 E_0}{\omega\mu} = \epsilon_0 \omega E_0$$

$$\text{or } \frac{\omega^2}{\beta^2} = \frac{1}{\epsilon_0\mu_0}. \tag{7.73}$$

But for a sinusoidal wave of the form

$$\mathbf{E} = E_0 \cos(\omega t - \beta z)\mathbf{i}$$

the **phase velocity** is determined by putting

$$\omega t - \beta z = \text{constant or } \frac{dz}{dt} = \frac{\omega}{\beta}.$$

But we have shown in the general theory that $1/\sqrt{(\mu_0 \epsilon_0)}$ is the speed of a plane electromagnetic wave in free space, so that (7.73) is valid and (7.62) is indeed satisfied by our calculated field $\mathbf{H}$.

(b) We know from the general theory that the amplitudes of the $\mathbf{E}$ and $\mathbf{H}$ vectors are linked in free space by

$$\frac{E_0}{H_0} = Z_0 = 120\pi \text{ ohms.}$$

Here, $H_0 = 200\pi/120\pi = 1.667$ A m^{-1}. The frequency of the sinusoidal oscillations is

$$f = \frac{v}{\lambda} = \frac{3 \times 10^8}{1.89} = 1.587 \times 10^8 \text{ Hz}$$

so the required angular frequency is

$$\omega = 2\pi f = 9.97 \times 10^8 \text{ rad s}^{-1}.$$

∎

7.10 Ocean Wave Interactions

There are several application areas of vector analysis, as we have seen thus far. In this section, we briefly present the application of vector analysis to the theory of ocean waves. Much of the research in this area relies heavily on vector analysis to establish results.

First order ocean waves can be regarded as the superposition of free and independent spectral components. However, a higher order, there exist interactions between spectral components which result in transfer of energy between the components[11]. The lowest order at which such interactions may occur involves the interaction of three waves, triad interactions. In such cases, two waves interact nonlinearly and transfer energy to a third component. Such interactions are important in finite depth conditions but are insignificant in deep water[4].

A general perturbation theory for the nonlinear resonant interaction of waves in a random sea was developed by Hasselmann[3,4]. He found that a set of four waves, called a quadruplet, could exchange energy when the following resonance conditions were satisfied:

$$\mathbf{k}_1 + \mathbf{k}_2 = \mathbf{k}_3 + \mathbf{k}_4 \tag{7.74}$$

$$\omega_1 + \omega_2 = \omega_3 + \omega_4 \tag{7.75}$$

where ω_j is the radian frequency and $\mathbf{k}_j$ the wavenumber vector ($j = 1, 2, 3, 4$). The wavenumber and frequency are related through the dispersion relationship $\omega^2 = gk \tanh kh$ for finite depth ocean and $\omega^2 = gk$ for infinite depth ocean. The four interacting wave components described by (7.74) and (7.75) form a quadruplet.

Hasselmann[4] described the nonlinear interactions between wave quadruplets in terms of their action density, n, where $n(\mathbf{k}) = Q(\mathbf{k})/\omega$ and $Q(\mathbf{k})$ is the wavenumber spectrum. He assumed that the waves were only weakly nonlinear and, hence, applied a perturbation technique to determine the rate of change in action density at $\mathbf{k}_1$ due to all quadruplet interactions involving $\mathbf{k}_1$. The rate of change $\partial n_1/\partial t$, which is directly related to $S_{nl}(f, \theta)$, is given by

$$
\begin{aligned}
\frac{\partial n_1}{\partial t} &= S_{nl} \\
&= \int \int \int G(\mathbf{k}_1, \mathbf{k}_2, \mathbf{k}_3, \mathbf{k}_4) \times \delta(\mathbf{k}_1 + \mathbf{k}_2 - \mathbf{k}_3 - \mathbf{k}_4) \\
&\quad \times \delta(\omega_1 + \omega_2 - \omega_3 - \omega_4) \\
&\quad \times [n_1 n_2(n_4 - n_2) + n_2 n_4(n_3 - n_1)] d\mathbf{k}_2 \, d\mathbf{k}_3 \, d\mathbf{k}_4
\end{aligned}
\tag{7.76}
$$

where $n_j = n(\mathbf{k}_j)$ is the action density at wavenumber $\mathbf{k}_j$ and G is a coupling coefficient.

The delta function in (7.76) ensures that the contribution to the integral only occurs for quadruplets that satisfy the resonance conditions (7.74) and (7.75). The integral equation (7.76) is primarily due to Boltzmann and attributed to his name as the Boltzmann integral for wind waves, in analogy to similar expressions used in theoretical physics, to describe the rate of change of particle density distributions in a system of interacting particles.

The resonant conditions define not only the frequencies of the spectral components that can interact nonlinearly but also their propagation directions, because (7.76) is a vector expression. This relationship can be represented in the form of a vector diagram as shown in Figure 7.13. To contribute

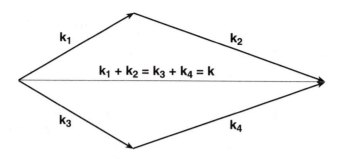

FIGURE 7.13
Wavenumber vectors satisfying resonance condition

to the integral expression (7.76), the frequencies associated with each of these wavenumbers must also be satisfied.

Hence, all components of the spectrum are potentially coupled and energy can be exchanged, not only between components of different frequencies, but also among components propagating different directions. The nonlinear energy transfer represented by (7.76) conserves both the total energy and momentum of the wave field, merely redistributing it among the spectrum. As a consequence of the symmetry of the resonance conditions with respect to the pairs of wavenumbers $(\mathbf{k}_1,\mathbf{k}_2)$ and $(\mathbf{k}_3,\mathbf{k}_4)$, the quadruplet interactions also conserve the wave action. The absolute value is equal for all wavenumbers within the quadruplet.

$$\frac{d\eta_1}{dt} = \frac{d\eta_2}{dt} = -\frac{d\eta_3}{dt} = -\frac{d\eta_4}{dt} \qquad (7.77)$$

This result is known as the property of detailed balance. In a more general form, detailed balance states that the absolute value of the change in action density, $|\Delta\eta_j|$, is the same for all components in a resonance set of wavenumbers.

7.11   Solution Techniques for S_{nl}

Numerical techniques have been proposed for solution of (7.76). The major goals of all these techniques have been to:

(a) improve computational speed;

(b) to provide a better physical understanding of the source term.

The significant computer time required for the solution arises for two reasons:

1. the fact that (7.76) represents a six dimensional integral;

2. the complexity of the complicated coupling coefficient, G.

It was suggested by many previous workers that it is important to first remove the delta functions by projecting the six-dimensional integral onto the three-dimensional resonance space defined (7.74) and (7.75). Although the details of the various projections differ, the goal was the same. The integral is then solved within the transformed space.

A conceptual difficulty arises in attempting to visualize the multi-dimensional integration space. It would add significantly to the understanding of the problem, if the wavenumber quadruplets which contribute significantly to the integral could be visualized in some simple form. Based on a suggestion by M. S. Longuet-Higgins[8], Hasselmann[3] proposed the so called *figures of eight* graphical representation of the interaction space. Figure 7.14 shows the interaction space for a given value $\mathbf{k} = \mathbf{k}_1 + \mathbf{k}_2 = \mathbf{k}_3 + \mathbf{k}_4$. Each of the curves is for a specific value of γ. For deep water conditions, where the dispersion relation $\omega^2 = gk\tanh kh$ reduces to $\omega^2 = gk$, resonance conditions (7.74) and (7.75) become

$$\mathbf{k}_1 + \mathbf{k}_2 = \mathbf{k}_3 + \mathbf{k}_4 = \mathbf{k} \tag{7.78}$$

$$\omega_1 + \omega_2 = \omega_3 + \omega_4 = \gamma\omega \tag{7.79}$$

where γ is a numerical constant. Equation (7.79) can be rewritten as

$$\sqrt{\mathbf{k}_1} + \sqrt{\mathbf{k}_2} = \sqrt{\mathbf{k}_3} + \sqrt{\mathbf{k}_4} = \gamma\sqrt{\mathbf{k}}. \tag{7.80}$$

Normalizing, such that $\mathbf{k} = 1$ and representing the angle between $\mathbf{k}_1$ and $\mathbf{k}_2$ as θ yields,

$$\cos\theta = \frac{1 + \mathbf{k}_1^2 - (\gamma - \sqrt{\mathbf{k}_1})^4}{2\mathbf{k}_1}. \tag{7.81}$$

For a fixed value of γ, the locus of all values of $\mathbf{k}_1$ that satisfy the resonance conditions is given by (7.80). The corresponding locus of $\mathbf{k}_2$ is given by (7.81). Typical integration curves for a given value of γ are illustrated in Figure 7.14.

A simple and perhaps the most straightforward representation of the integration space was presented by Webb[14]. For $\mathbf{k}_1$ and $\mathbf{k}_3$ fixed, the locus of the possible values of $\mathbf{k}_2$ and $\mathbf{k}_4$, which close the vector polygon represented by (7.74) and will also satisfy (7.75), trace out two egg-shaped figures. Figure 7.14 shows the loci of $\mathbf{k}_2$ and $\mathbf{k}_4$ for one combination of $\mathbf{k}_1$ and $\mathbf{k}_3$.

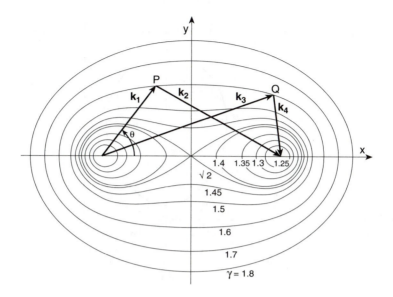

FIGURE 7.14
The so called *figure of eight* diagram

For this combination of $\mathbf{k}_1$ and $\mathbf{k}_3$, only wavenumber components that lie on these loci will satisfy the resonance conditions and contribute to the integral.

The direct integration techniques mentioned above provide valuable insight into the basic properties of the nonlinear source term S_{nl}. However, they were computationally too expensive for use in wave prediction models. As a result, approximation for narrow-peak spectra were proposed by Longuet-Higgins[8] and many other researchers.

The solution technique proposed by Webb[14] reduces the multi-dimensional integral represented by (7.76) to a series of line integrals around the **egg-shaped** loci resulting from the projection used to remove the delta functions. Although this operation simplifies the mathematical formulation, the number of possible loci along which line integrals need to be preformed is enormous. Figure 7.14 shows this loci for the values $\mathbf{k}_1 = 0.14$ rad/m, $\mathbf{k}_3 = 0.20$ rad/m. The $\mathbf{k}_1$ value is in the x-direction and $\mathbf{k}_4$ is 45 degrees clockwise. The solid line represents the loci of $\mathbf{k}_2$ while the dashed one represents that for $\mathbf{k}_4$.

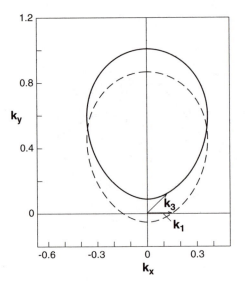

FIGURE 7.15
Loci of vectors

7.12 Graphical Simulation of a Vector Field

This section is concerned with the graphical simulation of a vector field. We will demonstrate here three plots of a vector field as illustrated in *The Mathematica* by Stephen Wolfram[15]. In the first simulation, we plot the vector field in two dimensions given by $\mathbf{A} = \mathbf{i} \sin x + \mathbf{j} \cos x + 0\mathbf{k}$. In the second simulation, we plot the gradient field of a scalar *potential* function such as the velocity potential given by $\mathbf{V} = \mathbf{i} \frac{\partial \phi}{\partial x} + \mathbf{j} \frac{\partial \phi}{\partial y} + 0\mathbf{k}$ in two dimensions where $\phi(x, y) = x^2 + y^2$.

The third example is the plot of the Hamiltonian vector field which are defined as $\frac{\partial \phi}{\partial y}$ and $-\frac{\partial \phi}{\partial x}$. Given a scalar-valued function ϕ on the $x - y$ plane, there are two common ways of constructing a vector field. The function's gradient vector field has first and second components given by $\frac{\partial \phi}{\partial x}$ and $\frac{\partial \phi}{\partial y}$, respectively. The two components of the function's Hamiltonian vector field are $\frac{\partial \phi}{\partial y}$ and $-\frac{\partial \phi}{\partial x}$.

The gradient field gives the direction in which the function is increasing most rapidly. If one treats the function as the Hamiltonian of a mechanical system, the Hamiltonian vector field gives the equation of motion in phase space. The package for plotting the vector field can be found in Mathematica as discussed in the text, *The Mathematica*. We simply use the Mathematica

FIGURE 7.16
Plot of a vector field

commands as illustrated by Wolfram. Interested readers are referred to Wolfram's work.

The following package plots two-dimensional vector fields. For doing vector field plots in three-dimensions, one uses the command

```
Graphics'PlotField3D'
```

before gaining access to the various plotting commands. The following is a summary of the commands that can now be used.

(1) `PlotVectorField[{phi_x, phi_y$}, {x, xmin, xmax},`
     `    {y, ymin,ymax}]`

This plots the vector field given by the vector-valued function in the range specified.

(2) `PlotGradientField[phi, {x, xmin, xmax},`
     `    {y, ymin, ymax}]`

This command plots the gradient vector field of the scalar-valued function ϕ.

(3) `PlotHamiltonianField[phi, {x, xmin, xmax},`
     `    {y, ymin, ymax}]`

This command plots the Hamiltonian vector field of the scalar-valued function ϕ.

1. Plotting Vector Fields in Two-Dimensions

(a) We can load the package with the following command at the Mathematica prompt:

$$In[1] := < Graphics'PlotField'$$

Once the package is loaded, then the vector field can be plotted or manipulated as desired.

(b) As an example, we can give the two components of the vector field are given by $\sin[x]$ and $\cos[y]$.

$$In[2] := PlotVectorField[sin[x], cos[y]\},$$
$$\{x, 0, pi\},\{y, 0, pi\}]$$

2. PlotGradientField and PlotHamiltonianField

These commands compute the formulae for the partial derivatives of the scalar-valued function given as an argument. This means that the functions one gives must be such that Mathematica can compute their derivatives, i.e., continuously differentiable functions.

(a) Here is the gradient field of the potential

$$\phi(x, y) = x^2 + y^2.$$

$$In[3] := PlotGradientField[x^2 + y^2,$$
$$\{x, -3, 3\},\{y, -3, 3\}]$$

(b) In the plane, the gradient and Hamiltonian vector fields of the same function are always orthogonal.

$$In[4] := PlotHamiltonianField[x^2 + y^2,$$
$$\{x, -3, 3\},\{y, -3, 3\}]$$

Exercises

1. What is the potential function for a central force field in which the attraction on a particle varies as the square of the distance from the origin? Inversely as the distance from the origin?

2. What is the potential function for the gravitational field of a uniform circular disk at any point on the axis of the disk?

FIGURE 7.17
Gradient field plot of $\phi = x^2 + y^2$

FIGURE 7.18
Hamiltonian vector field

3. What is the potential function for the gravitational field of a uniform sphere of radius a and mass M? Show that the attraction of the sphere at a point P a distance r from the center of the sphere is

$$\mathbf{F} = \begin{cases} -\frac{M\mathbf{R}}{a^3} & r \leq a \\ -\frac{M\mathbf{R}}{r^3} & r \geq a \end{cases}$$

4. If a conductor is defined to be a body in whose interior the electric field is everywhere zero, show that any charge on a conductor must be located entirely on its surface.

5. In cylindrical coordinates, the electric flux density vector $\mathbf{D}$ in a region is given by

$$\mathbf{D} = 4\rho^2 \boldsymbol{\rho} \, \mathrm{C \ m}^{-2}$$

Verify that the total flux of $\mathbf{D}$ over the surface of the cylinder $\rho = 2, |z| < 5$ is equal to the total charge inside the cylinder.

6. Calculate the total magnetic flux crossing the $z = 0$ plane for $\rho \leq 5 \times 10^{-2}$ m when

$$\mathbf{B} = \frac{0.2}{\rho} \sin^2 \phi \mathbf{k} \ \mathrm{Wb \ m}^{-2}$$

7. Show that the fields

$$\mathbf{E} = E_0 \sin x \sin t \mathbf{j} \ \text{and} \ \mathbf{H} = \frac{E_0}{\mu_0} \cos x \cos t \mathbf{k}$$

satisfy only three of the four Maxwell equations and thus are both valid solutions.

8. In free space, a time-dependent magnetic vector potential $\mathbf{A} = exp(-2\rho)\mathbf{k}$ is present. Calculate expressions for the vector fields $\mathbf{B}$, $\mathbf{H}$ and $\mathbf{J}$. Calculate also the total current crossing the $z = 0$ plane between $\rho = 0$ and $\rho = 10$ mm.

9. An electric charge of 3 C is moved from the point $P_1(1, 2, 0)$ to the point P_2 (2, 8, 0) (distances are in meters). Find the work done if the electric field is

$$\mathbf{E} = 4xy\mathbf{i} + 2x^2\mathbf{j} \ \mathrm{V \ m}^{-1}$$

   (a) by direct integration, and

(b) by finding the associated scalar potential ϕ.

10. Electric charge is distributed within a spherical region of radius a such that the electric field is

$$\mathbf{E} = Ar^3\mathbf{r} \text{ V m}^{-1}$$

where A is a constant. Calculate

(a) the total charge within the sphere,
(b) the total electrostatic energy stored within this region. (Note: $\frac{1}{2}\epsilon_0|\mathbf{E}|^2$ is the electrostatic energy density.)

11. The fields in a spherical wave at a large distance r from a dipole antenna in free space, using spherical coordinates,

$$\mathbf{E} = \frac{150}{r}\sin\theta\cos(\omega t - \beta r)\boldsymbol{\theta} \text{ V m}^{-1}$$
$$\mathbf{H} = \frac{0.2}{r}\sin\theta\cos(\omega t - \beta r)\boldsymbol{\phi} \text{ A m}^{-1}$$

calculate the instantaneous and average powers crossing a hemisphere $0 < \theta < \pi/2$ at a distance of 1 km from the antenna.

12. Show, by substitution, that the fields

$$\mathbf{E} = E_0\cos(\omega t + \beta z)\mathbf{i} \text{ and } \mathbf{H} = -\frac{E_0}{Z}\cos(\omega t + \beta z)\mathbf{j}$$

(where $Z = \sqrt{(\mu/\epsilon)}$) satisfy Maxwell's four equations for a perfect charge-free dielectric characterized by μ and ϵ.

13. The displacement current density in a perfect dielectric with $\epsilon = 4\epsilon_0$ and $\mu = 5\mu_0$ is given by

$$\frac{\partial \mathbf{D}}{\partial t} = 2\cos(\omega t - 5z)\mathbf{i} \ \mu\text{A m}^{-2}$$

Find expressions for the fields $\mathbf{D}$, $\mathbf{E}$, $\mathbf{B}$ and $\mathbf{H}$, ignoring all constant terms. Recalculate $\partial\mathbf{D}/\partial t$, and hence find the value of ω.

14. In free space, an electromagnetic wave has an $\mathbf{E}$ vector

$$\mathbf{E}(z, t) = 150\sin(\omega t - \beta z)\mathbf{i} \text{ V m}^{-1}$$

Calculate the total average power passing through a rectangular area of sides 30 mm and 15mm in the $z = 0$ plane.

15. In a cylindrical conductor or radius 2 mm, the current density $\mathbf{J}$ is given by

$$\mathbf{J} = 10^3 e^{-400\rho} \mathbf{k} \text{ A m}^{-2}$$

Calculate the total current I in the conductor.

16. In spherical coordinates, a magnetic field $\mathbf{H}$ is given by

$$\mathbf{H} = 10^6 r \sin\theta\, \boldsymbol{\phi} \text{ A m}^{-1}$$

Deduce the r-component of $\nabla \times \mathbf{H}$, and hence the current flowing across the spherical "cap"

$$r = 1\text{mm}, \quad 0 < \theta < \pi/6, \quad 0 < \phi < 2\pi$$

17. Show directly from Maxwell's equations that the charge density ρ and the electric current density $\mathbf{j}$ obey the conservation law

$$\frac{\partial \rho}{\partial t} + \nabla \cdot \mathbf{j} = 0.$$

18. Derive the formula for the electric field $\mathbf{E}$ due to a point charge Q using Gauss' law.

19. Show from Maxwell's equations in a vacuum that the magnetic field $\mathbf{B}$ obeys the wave equation.

20. For the electromagnetic wave in which the electric field is given by $\mathbf{E} = \mathbf{E}_0 f(\mathbf{k} \cdot \mathbf{x} - \omega t)$, calculate the corresponding magnetic field $\mathbf{B}$. What can be deduced about the directions of the vectors $\mathbf{E}$ and $\mathbf{B}$?

21. Using Maxwell's equations in a vacuum, obtain an equation in the form of a conservation law for the rate of change of the energy $w = |\mathbf{B}|^2/2 + |\mathbf{E}|^2/2c^2$ of an electromagnetic wave.

References

[1] Arfken, G. (1985). *Mathematical Methods for Physicists*, Academic Press, Inc., Orlando.

[2] Curle, N. (1968). *Modern Fluid Dynamics*, **Vol. 1**, Van Nostrand, London.

[3] Hasselmann, K. (1962), On the nonlinear energy transfer in a gravity spectrum, Part 1: General Theory, *J. Fluid Mech.*, **12**, 481-500.

[4] Hasselmann, K. (1963), On the nonlinear energy transfer in a gravity spectrum, Part 2: Conservation theorems, wave particle analogy, irreversibility, *J. Fluid Mech.*, **15**, 385-398.

[5] Kreyszig, E. (1988). *Advanced Engineering Mathematics*, John Wiley & Sons, New York.

[6] Lewis, P.E. and Ward, J.W. (1989). *Vector Analysis for Engineer and Scientists*, Addison-Wesley Publishing, New York.

[7] Lighthill, J. (1986). *An Informal Introduction to Theoretical Fluid Mechanics*, Oxford University Press, U.K.

[8] Longuet-Higgins, M.S. (1976). On the nonlinear transfer of energy in the peak of a gravity wave spectrum: a simplified model, *Proc. Roy. Soc. London*, **A347**, 311-328.

[9] Marsden, J.E. and Tromba, A.J. (1988). *Vector Calculus*, W.H. Freeman, New York.

[10] Matthews, P.C. (1998). *Vector Calculus*, Springer-Verlag, London, U.K.

[11] Phillips, O.M. (1960). *The Dynamics of Upper Oceans*, Cambridge University Press, U.K.

[12] Rahman, M. (1995). *Water Waves: Relating Modern Theory with Advanced Engineering Applications*, Oxford University Press, U.K.

[13] Trim, D. (1998). *Calculus for Engineers*, Prentice Hall Canada, Inc., Scarborough, Ontario.

[14] Webb, D.J. (1978). Nonlinear transfer between sea waves, *Deep-Sea Res.*, **25**, 279-298.

[15] Wolfram, S. (1996). *The Mathematica*, Cambridge University Press, U.K.

[16] Wylie, C.R. (1975).*Advanced Engineering Mathematics*, McGraw-Hill, New York.

[17] Zill, D.G. and Cullen, M.R. (1992). *Advanced Engineering Mathematics*, PWS–Kent, Boston.

Appendix A

Answers to Odd-Numbered Exercises

Chapter 2

3.

$$\mathbf{A} \cdot \mathbf{B} = |\mathbf{A}||\mathbf{B}| \cos \theta$$
$$\cos \theta = \frac{\mathbf{A} \cdot \mathbf{B}}{|\mathbf{A}||\mathbf{B}|}$$
$$= \frac{(2\mathbf{i} - 3\mathbf{j} + \mathbf{k}) \cdot (3\mathbf{i} - \mathbf{j} - 2\mathbf{k})}{(\sqrt{14})(\sqrt{14})}$$
$$= \frac{7}{14}$$
$$\cos \theta = \frac{7}{14}$$

7. $-\frac{3}{\sqrt{10}}\mathbf{i} + \frac{1}{\sqrt{10}}\mathbf{j}$ and $\frac{3}{\sqrt{10}}\mathbf{i} - \frac{1}{\sqrt{10}}\mathbf{j}$

11. $\frac{9}{\sqrt{17}}$

13. (b) $\frac{13}{5}$

15. $\sqrt{4014}$

17. (a) 10 (b) 50

19. (a) $3x - 2y + 4z - 11 = 0$

  (b) $20x + 14y - 8z = 24$

  (c) $4x + y + 2z = 12$

Chapter 3

1. (a) $\mathbf{F}'(t) = \cos t\mathbf{i} - \sin t\mathbf{j}$
    (b) True since $\mathbf{k} \cdot \mathbf{F}'(t) = 0$
    (c) $t = n\pi, (n = 0, \pm 1, \pm 2, \ldots)$
    (d) Yes, $\sqrt{2}$
    (e) Yes, 1
    (f) $\mathbf{F}''(t) = -\sin t\mathbf{i} - \cos t\mathbf{j}$

9. (a) $v = 3/2$ (b) $\mathbf{a}(t) = -\cos t(\mathbf{i} - \mathbf{j}) - \sin t(\mathbf{i} + \mathbf{j})$
    (c) $-2/3 \sin t(\mathbf{i} - \mathbf{j}) + 2/3 \cos t(\mathbf{i} + \mathbf{j}) + 1/3\mathbf{k}$ (d) $\kappa = \frac{8\sqrt{5}}{27}$

11. (a) $\mathbf{T}(t) = \frac{1}{(1+t^2)}(1, \sqrt{2}t, t^2), \mathbf{N}(t) = \frac{1}{\sqrt{2}(1+t^2)}(-\sqrt{2}t, \sqrt{t}(1 - t^2), 2t)$
    (b) $\kappa = \frac{1}{3\sqrt{2}(1+t^2)^2}$

13. (a) $\mathbf{T}(t) = \frac{1}{(2t^2+1)}(2t^2, 2t, 1), \mathbf{N}(t) = \frac{1}{(2t^2+1)}(2t, 1 - 2t^2, -2t)$
    (b) $\kappa = \frac{2t}{(2t^2+2)^2}$

15. (a) The normal plane is $y - 6x = \pi$, osculating plane is $x + 6y = 6\pi$
    (b) The normal plane is $x + 2y + 3z = 6$, osculating plane is $3x - 3y + z = 1$

17. The speed of the object is $|\mathbf{v}(t)| = \sqrt{25t^2 + 2}$ while its position is $\mathbf{r}(t) = t\mathbf{i} - t\mathbf{j} + 5/2t^2\mathbf{k}$

19. (a) C intersects the xz-plane where $y = 0$, the point is $(15/8, 0, -\ln 2)$
    (b) The tangent line is $x = 1 - 3s, y = 1 + 2s, z = s$
    (c) The equation of the normal plane is $3x - 2y - z = 1$

Chapter 4

1. (a) $7\sqrt{3} - 16$ (b) $-672\sqrt{2}$ (c) 1 (d) $-1/2 - \sqrt{3}$

3. (a) $\nabla f(x, y) = e^x \sin y\mathbf{i} + e^x \cos y\mathbf{j}$
   (b) $\nabla f(1, \pi/4) = \frac{1}{\sqrt{2}}e(\mathbf{i} + \mathbf{j})$
   (c) $\nabla f(1, \pi/4) \cdot \mathbf{u} = \frac{1}{\sqrt{10}}e$

5. (a) $\nabla f(x, y, z) = (y + z^3, x + z^2, 2yz + 3xz^2)$
   (b) $\nabla f(2, 0, 3) = (27, 11, 54)$
   (c) $\nabla f(2, 0, 3) \cdot \mathbf{u} = 43/3$

7. (a) $\nabla f(1,0) = (1,2)$ is the direction and the maximum rate is

$$|\nabla f(1,0)| = \sqrt{5}$$

(b) The maximum rate of change is $|\nabla f(1,2)| = 2\sqrt{5}/5$ in the direction $(2/5, 4/5)$ or $(2,4)$

(c) The maximum rate of change is $|\nabla f(4,10)| = \sqrt{17}/6$ in the direction $(2/3, 1/6)$ or $(4,1)$

(d) Maximum rate of change is $|\nabla f(4,3,-1)| = \sqrt{11}$ in the direction $(1,-1,-3)$

(e) The maximum rate of change is $|\nabla f(\pi/6, -\pi/8)| = \sqrt{13/2}$ in the direction of $(-3\sqrt{2}/2, -\sqrt{2})$ or $(-3,-2)$

(f) The maximum rate of change is $|\nabla f(4,2,1)| = \sqrt{17}/2$ in the direction $(1/2, 0, -2)$ or $(1, 0, -4)$

9. (a) Proceed in the direction $\nabla f(60, 100) = (-1, 2, -4)$

(b) The maximum slope is equal to the maximum directional derivative, which is $|(1, 2, -4)| = \sqrt{17.44}$ and

$$\theta = \tan^{-1}\sqrt{17.44} \approx 76.5 \text{ degrees}$$

11. (a) Satisfies (b) Satisfies (c) Satisfies provided that $p^2 = q^2$

13. $40[(z-y)\mathbf{i} + (x-z)\mathbf{j} + (y-x)\mathbf{k}]$

15. (a) $2\mathbf{R} \cdot \mathbf{A}$ (b) $\mathbf{R} \times \mathbf{A}$ (c) $2r^2\mathbf{A}$ (d) $4(\mathbf{A} \cdot \mathbf{R})^3\mathbf{A}$

(e) $(\mathbf{A} \cdot \mathbf{R})/r$ (f) $(\mathbf{A} \cdot \mathbf{R})\mathbf{A}$ (g) 0 (h) $2\mathbf{A}$ (i) 6

23. $-\frac{GMm}{R^3}(z^2 - \frac{1}{2}x^2 - \frac{1}{2}y^2)$

Chapter 5

1. (a) $\int_C x\,ds = \int_0^1 t^3\sqrt{9t^4 + 1}\,dt = \frac{1}{54}(10^{3/2} - 1)$

(b) $\int_C xy^4\,ds = \int_{-\pi/2}^{\pi/2}[(4)^5\cos t\sin^4 t]4\,dt = 1638.4$

(c) $\int_C(x - 2y^2)\,dy = \int_{-2}^1(2x^2 - 4x^5)\,dx = 48$

(d) $C = C_1 + C_2$ On C_1 : $x = x, y = 0, 0 \le x \le 2$. On C_2 : $x = x, y = 2x - 4, 2 \le x \le 3$. Then

$$\int_C xy dx + (x - y)dy$$

$$= \int_{C_1} xy dx + (x - y)dy + \int_{C_2} xy dx + (x - y)dy$$

$$= \int_0^2 0 dx + \int_2^3 [(2x^2 - 4x) + (4 - x)(2)]dx$$

$$= \int_2^3 (2x^2 - 6x + 8)dx = \frac{17}{3}$$

(e)

$$\int_C x\sqrt{y} dx + 2y\sqrt{x} dy$$

$$= \int_{C_1} x\sqrt{y} dx + 2y\sqrt{x} dy + \int_{C_2} x\sqrt{y} dx + 2y\sqrt{x} dy$$

$$= \int_0^{\pi/2} [-\cos t(\sin t)^{3/2} + 2\sin t(\cos t)^{3/2}]dt$$

$$+ \int_0^1 [16t\sqrt{2t + 1} + 8(2t + 1)\sqrt{t}]dt$$

$$= \frac{32\sqrt{3} + 66}{5}$$

3. 0

5. (a) 2a (b) 0

7. $\int_c \mathbf{F} \cdot d\mathbf{r} = \int_0^1 (2te^{t^2-1} + 3t^7)dt = 11/8 - 1/e$

9. $m = \int_0^{2\pi} (t^2 + 1)\sqrt{2} dt = \sqrt{2}(8/3\pi^3 + 2\pi)$.
$\bar{x} = \frac{1}{\sqrt{2}(8/3\pi^3+2\pi)} \int_0^{2\pi} \sqrt{2}(t^3 + t)dt = \frac{3\pi(2\pi^2+1)}{4\pi^2+3}$
$\bar{y} = 0, \bar{z} = 0$. Hence, we have $(\bar{x}, \bar{y}, \bar{z}) = (\frac{3\pi(2\pi^2+1)}{4\pi^2+3}, 0, 0)$

11. Here $\mathbf{F}(x, y) = (2x \sin y)\mathbf{i} + (x^2 \cos y - 3y^2)\mathbf{j}$. Then $f(x, y) = x^2 \sin y - y^3$ is a potential for $\mathbf{F}$ i.e., $\nabla f = \mathbf{F}$ so $\mathbf{F}$ is conservative and thus the line integral is independent of path. Therefore,

$$\int_C 2x \sin y dx + (x^2 \cos y - 3y^2)dy = f(5, 1) - f(-1, 0)$$

$$= 25 \sin(1) - 1$$

13. (a) $P = -\frac{y}{x^2+y^2}$, $\frac{\partial P}{\partial y} = \frac{y^2-x^2}{(x^2+y^2)^2}$ and $Q = \frac{x}{x^2+y^2}$ $\frac{\partial Q}{\partial x} = \frac{y^2-x^2}{(x^2+y^2)^2}$.
Thus, $\frac{\partial P}{\partial y} = \frac{\partial Q}{\partial x}$

(b) C_1 : $x = \cos t$, $y = \sin t$, $0 \leq t \leq \pi$; C_2 : $x = \cos t$, $y = \sin t$, $t = 2\pi$ to $t = \pi$. Then

$$\int_{C_1} \mathbf{F} \cdot d\mathbf{r} = \int_0^\pi \frac{(-\sin t)(-\sin t) + (\cos t)(\cos t)}{\cos^2 t + \sin^2 t} dt = \pi \text{ and}$$

$$\int_{C_2} \mathbf{F} \cdot d\mathbf{r} = \int_{2\pi}^\pi dt = -\pi$$

Because these aren't equal, the line integral of $\mathbf{F}$ is not independent of path. This does not contradict Theorem 5.3 since the domain of $\mathbf{F}$ (which is $\mathcal{R}^2$ except the origin) is not simply connected.

15. (a) $\int\int_S y ds = \int_0^3 \int_0^{(6-2y/3)} y\sqrt{14}\,dx\,dy = 3\sqrt{14}$

(b) $\int\int_S xz ds = \int_0^1 \int_0^{1-y}(x(1-y) - x^2)\sqrt{3}\,dx\,dy = \frac{1}{8\sqrt{3}}$

(c) $\int\int_S x ds = \int_0^2 \int_0^2 x\sqrt{4x^2 + 17}\,dx\,dz = \frac{3\sqrt{33}-17\sqrt{17}}{6}$

(d)

$$\int\int_S (y^2 + z^2) ds = \int_0^{2\pi} \int_0^2 r^2\sqrt{4r^2 + 1}\,r\,dr\,d\theta$$

$$= \frac{\pi}{60}(391\sqrt{17} + 1)$$

17. (a) $2e^{-10}e^{-2}$ (b) 0 (c) $\frac{2\pi}{3}$ (d) -2π (e) $\frac{10\pi}{3}$ (f) $\frac{\pi\sqrt{2}}{4}$

19. $2\pi(1 - e^{-2})$

21. (a) $\frac{\pi(8\sqrt{2}-7)}{6}$ (b) 4π

23. (a) $\frac{\pi(2-\sqrt{2})}{3}$ (b) $\frac{(4\tan^{-1}2-\pi)}{12}$ (c) $\frac{32\pi}{3}$

Chapter 6

1. (a) (i) $\oint_C x^2 y dx + xy^3 dy = -\frac{1}{12}$

(ii) $\oint_C x^2 y dx + xy^3 dy = \int_0^1 \int_0^1 (y^3 - x^2)dx\,dy = -\frac{1}{12}$

(b) (i) $\oint_C x dx - x^2 y^2 dy = -\frac{1}{5}$

(ii) $\oint_C x dx - x^2 y^2 dy = \int_0^1 \int_x^1 (-2xy^2)dy\,dx = -\frac{1}{5}$

(c) (i) $\oint_C (x + 2y)dx + (x - 2y)dy = -\frac{1}{6}$

(ii) $\oint_C (x + 2y)dx + (x - 2y)dy = \int_0^1 \int_{x^2}^x (1 - 2)dy\,dx = -\frac{1}{6}$

(d) (i) $\oint_C (x^2 + y^2)dx + 2xydy = 0$

   (ii) $\oint_C (x^2 + y^2)dx + 2xydy = \int \int_D (2y - 2)dA = 0$

3. By Green's Theorem, $W = \int_C \mathbf{F} \cdot d\mathbf{r} = \int_C x(x + y)dx + xy^2 dy = \int \int_D (y^2 - x)dydx$ where C is the path described in the question and D is the triangle bounded by C. So
$W = \int_0^1 \int_0^1 (y^2 - x)dydx = -\frac{1}{12}$

5. (a) $\int \int_S \nabla \times \mathbf{F} \cdot d\mathbf{S} = \oint_C \mathbf{F} \cdot d\mathbf{r} = \int_0^{2\pi} \cos^2 t \, dt = \pi$

   (b) $\int \int_S \nabla \times \mathbf{F} \cdot d\mathbf{S} = \oint_C \mathbf{F} \cdot d\mathbf{r} = \int_0^{2\pi} (\cos^2 t - \sin^3 t)dt = \pi$

   (c) $\int \int_S \nabla \times \mathbf{F} \cdot d\mathbf{S} = \oint_C \mathbf{F} \cdot d\mathbf{r} = \int_0^{2\pi} \cos^4 t \, dt = \frac{3\pi}{4}$

   (d) $\int \int_S \nabla \times \mathbf{F} \cdot d\mathbf{S} = \oint_C \mathbf{F} \cdot d\mathbf{r} = 2 \int_0^{2\pi} (-1 + \cos 2t + \sin 2t)dt = -4\pi$

7. $\int_C \mathbf{F} \cdot d\mathbf{r} = \int_S \nabla \mathbf{F} \cdot \mathbf{n} dS = \int_0^1 \int_0^{1-x} (8 - 6x - 10y)dydx = \frac{4}{3}$

9. $\int_C \mathbf{F} \cdot d\mathbf{r} = \int_S \nabla \mathbf{F} \cdot d\mathbf{S} = \int \int_D 0 dA = 0$

11. $\int_C \mathbf{F} \cdot d\mathbf{r} = \int \int_S \nabla \times d\mathbf{S} = 2 \int_0^{2\pi} \int_0^1 r^3 dr d\theta = \pi$

13. $\int \int_S \nabla \times \mathbf{F} \cdot d\mathbf{S} = 16$

15. $\int \int_S \mathbf{F} \cdot d\mathbf{S} = \int \int \int_E (-z - z + 2z)dV = 0$

17. $\int \int_S \mathbf{F} \cdot d\mathbf{S} = \int \int \int_E (1 - 1)dV = 0$

19. $\int \int_S \mathbf{F} \cdot d\mathbf{S} = \int_0^{2\pi} \int_0^{\pi/2} \int_1^2 3(\rho^2 \sin^2 \phi + 1)\rho^2 \sin \phi d\rho d\phi d\theta = \frac{194}{5}\pi$

21. $\int \int_S \mathbf{F} \cdot d\mathbf{S} = \int_0^{2\pi} \int_0^{\pi} \int_2^3 (4\rho^2 \sin^2 \phi \cos^2 \theta)(\rho^2 \sin \phi)d\rho d\phi d\theta = \frac{3378}{15}\pi$

23. Here the same idea used in question 22 is pursued. $S_2 = S \cup S_1$, where S is the part of the paraboloid $x^2 + y^2 + z = 2$ that lies above the plane $z = 1$, and S_1 is the disk $x^2 + y^2 = 1$ on the plane $z = 1$ oriented downward, and then we apply the Divergence Theorem. So we have
$\int \int_{S_1} \mathbf{F} \cdot d\mathbf{S} = \int \int_{S_1} \mathbf{F} \cdot \mathbf{n} dS = \int \int_{S_1} (-1)dS = -\pi$. Then
$\int \int_{S_2} \mathbf{F} \cdot d\mathbf{S} = \int_0^1 \int_0^{2\pi} \int_1^{2-r^2} r dz d\theta dr = \frac{\pi}{2}$. Thus, the flux of $\mathbf{F}$ across S is

$$\int \int_S \mathbf{F} \cdot d\mathbf{S} = \int \int_{S_2} \mathbf{F} \cdot d\mathbf{S} - \int \int_{S_1} \mathbf{F} \cdot d\mathbf{S} = \frac{3\pi}{2}$$

25. $\int \int_S (2x + 2y + z^2)dS = \int \int \int_B dV = \frac{4}{3}\pi$.

27. $\int \int_S \nabla \times \mathbf{F} \cdot d\mathbf{S} = \oint_C \mathbf{F} \cdot d\mathbf{r} = \int_0^{2\pi} (16 \cos^3 t \sin t + 4 \sin^2 t)dt = 4\pi$

29. $\int\int_S \mathbf{F} \cdot d\mathbf{S} = \int_0^{2\pi} \int_0^1 \int_0^2 (3r^2 + 3z^2) r\,dz\,dr\,d\theta = 11\pi$

31. Let S_1 be the sphere of radius 1 with center at the origin and outer normal $\mathbf{n}_1$. Let S_2 be the surface of the ellipsoid with outer unit normal $\mathbf{n}_2$ and let E be the solid region between S_1 and S_2. Then the outward flux of $\mathbf{F}$ through the ellipsoid is given by

$$\int\int_{S_2} \mathbf{F} \cdot \mathbf{n}_2 dS = \int\int_{S_1} dS + \int\int\int_E 0\,dV$$
$$= \text{(surface area of the unit sphere)}$$
$$= 4\pi$$

Chapter 7

17. We have

$$\frac{\partial \rho}{\partial t} = \varepsilon_0 \nabla \cdot \frac{\partial \mathbf{E}}{\partial t}$$
$$= \varepsilon_0 \nabla \cdot (\nabla \times \mathbf{B} - \mu_0 \mathbf{j}) / \varepsilon_0 \mu_0$$
$$= -\nabla \cdot \mathbf{j}$$

because the combination div curl is always zero.

19. Taking the curl, we have

$$\nabla \times (\nabla \times \mathbf{B}) = \mu_0 \varepsilon_0 \frac{\partial \nabla \times \mathbf{E}}{\partial t}.$$

Expanding the right hand side, we obtain

$$-\nabla^2 \mathbf{B} = -\mu_0 \varepsilon_0 \frac{\partial^2 \mathbf{B}}{\partial t^2},$$

so $\mathbf{B}$ obeys exactly the same wave equation as $\mathbf{E}$.

21. The energy can be written $w = \mathbf{B} \cdot \mathbf{B}/2 + \mathbf{E} \cdot \mathbf{E}/2c^2$, so the rate of change of energy is

$$\frac{\partial w}{\partial t} = \mathbf{B} \cdot \frac{\partial \mathbf{B}}{\partial t} + \frac{1}{c^2} \mathbf{E} \cdot \frac{\partial \mathbf{E}}{\partial t}$$
$$= -\mathbf{B} \cdot \nabla \times \mathbf{E} + \mathbf{E} \cdot \nabla \times \mathbf{B}$$
$$= -\nabla \cdot (\mathbf{E} \times \mathbf{B}).$$

This gives the conservation law

$$\frac{\partial w}{\partial t} + \nabla \cdot \mathbf{P} = 0$$

where $\mathbf{P} = \mathbf{E} \times \mathbf{B}$ is known as the Poynting vector, representing the energy flux of the electromagnetic wave.

Appendix B

Vector Formulae at a glance

Vector algebra

$$\mathbf{A} \cdot \mathbf{B} = \mathbf{B} \cdot \mathbf{A}$$
$$\mathbf{A} \times \mathbf{B} = -\mathbf{B} \times \mathbf{A}$$
$$\mathbf{A} \cdot (\mathbf{B} \times \mathbf{C}) = \mathbf{B} \cdot (\mathbf{C} \times \mathbf{A}) = \mathbf{C} \cdot (\mathbf{A} \times \mathbf{B})$$
$$\mathbf{A} \times (\mathbf{B} \times \mathbf{C}) = (\mathbf{A} \cdot \mathbf{C})\mathbf{B} - (\mathbf{A} \cdot \mathbf{B})\mathbf{C}.$$

Gradient

$$\nabla(\Phi\theta) = \Phi\nabla\theta + \theta\nabla\Phi$$
$$\nabla(\mathbf{A} \cdot \mathbf{B}) = (\mathbf{A} \cdot \nabla)\mathbf{B} + (\mathbf{B} \cdot \nabla)\mathbf{A} + \mathbf{A} \times (\nabla \times \mathbf{B}) + \mathbf{B} \times (\nabla \times \mathbf{A}).$$

Divergence

$$\nabla \cdot (\Phi\mathbf{A}) = \Phi\nabla \cdot \mathbf{A} + \mathbf{A} \cdot \nabla\Phi$$
$$\nabla \cdot (\mathbf{A} \times \mathbf{B}) = -\mathbf{A} \cdot (\nabla \times \mathbf{B}) + \mathbf{B} \cdot (\nabla \times \mathbf{A}).$$

where ϕ is a scalar function.

Curl

$$\nabla \cdot (\nabla\Phi) = \nabla^2\Phi$$
$$\nabla \cdot (\nabla \times \mathbf{A}) = 0$$
$$\nabla \times (\nabla\Phi) = 0$$
$$\nabla \times (\nabla \times \mathbf{A}) = \nabla(\nabla \cdot \mathbf{A}) - \nabla^2\mathbf{A}.$$

Unit Tangent, Normal, Binormal and the Osculating Plane

$$\mathbf{R} = x\mathbf{i} + y\mathbf{j} + z\mathbf{k} \text{ the position vector}$$

$$\mathbf{T} = \frac{\dot{\mathbf{R}}}{|\dot{\mathbf{R}}|} \text{ unit tangent}$$

$$\mathbf{N} = \frac{\dot{\mathbf{T}}}{|\dot{\mathbf{T}}|} \text{ unit normal}$$

$$\mathbf{B} = \mathbf{T} \times \mathbf{N} \text{ binormal}$$

$$\vec{P_1 P} \cdot \left\{ \frac{d\mathbf{R}}{dt} \times \frac{d^2\mathbf{R}}{dt^2} \right\}_{t_0} = 0 \text{ osculating plane}$$

where t_0 is a given point.

Curvature, Tangential and Normal acceleration

$$\mathbf{R} = x\mathbf{i} + y\mathbf{j} + z\mathbf{k} \text{ the position vector}$$

$$a_T = \frac{\dot{\mathbf{R}} \cdot \ddot{\mathbf{R}}}{|\dot{\mathbf{R}}|} \text{ tangential acceleration}$$

$$a_N = \frac{|\dot{\mathbf{R}} \times \ddot{\mathbf{R}}|}{|\dot{\mathbf{R}}|} \text{ normal acceleration}$$

$$\kappa = \frac{|\dot{\mathbf{R}} \times \ddot{\mathbf{R}}|}{|\dot{\mathbf{R}}|^3}.$$

Serret–Frenet Equations

$$\tau = \left| \frac{d\mathbf{B}}{ds} \right|, \quad \text{torsion}$$

$$\frac{d\mathbf{T}}{ds} = \kappa\mathbf{N}$$

$$\frac{d\mathbf{B}}{ds} = -\tau\mathbf{N}$$

$$\frac{d\mathbf{N}}{ds} = \tau\mathbf{B} - \kappa\mathbf{T}.$$

Green's Theorem

$$\oint_C P\,dx + Q\,dy = \int\int_R \left(\frac{\partial Q}{\partial x} - \frac{\partial P}{\partial y} \right) dx\,dy.$$

Divergence Theorem

$$\int\int_S \mathbf{A} \cdot \mathbf{n} \; dS = \int\int\int_V \nabla \cdot \mathbf{A} \; dV$$

where V is a volume bounded by a closed surface S

Stokes' Theorem

$$\oint_C \mathbf{A} \cdot d\mathbf{R} = \int\int_S (\nabla \times \mathbf{A}) \cdot \mathbf{n} \; dS$$

where S is an open surface bounded by a closed contour C

Cartesian Coordinates (x, y, z)

$$\nabla G = \mathbf{i}\frac{\partial G}{\partial x} + \mathbf{j}\frac{\partial G}{\partial y} + \mathbf{k}\frac{\partial G}{\partial z}$$

$$\nabla \cdot \mathbf{A} = \frac{\partial A_x}{\partial x} + \frac{\partial A_y}{\partial y} + \frac{\partial A_z}{\partial z}$$

$$\nabla \times \mathbf{A} = \mathbf{i}\left[\frac{\partial A_z}{\partial y} - \frac{\partial A_y}{\partial z}\right] + \mathbf{j}\left[\frac{\partial A_x}{\partial z} - \frac{\partial A_z}{\partial x}\right]$$

$$+ \mathbf{k}\left[\frac{\partial A_y}{\partial x} - \frac{\partial A_x}{\partial y}\right]$$

$$\nabla^2 G = \frac{\partial^2 G}{\partial x^2} + \frac{\partial^2 G}{\partial y^2} + \frac{\partial^2 G}{\partial z^2}.$$

Cylindrical Coordinates (r, θ, z)

$$\nabla G = \mathbf{r}\frac{\partial G}{\partial r} + \frac{1}{r}\boldsymbol{\theta}\frac{\partial G}{\partial \theta} + \mathbf{k}\frac{\partial G}{\partial z}$$

$$\nabla \cdot \mathbf{A} = \frac{1}{r}\frac{\partial(rA_r)}{\partial r} + \frac{1}{r}\frac{\partial A_\theta}{\partial \theta} + \frac{\partial A_z}{\partial z}$$

$$\nabla \times \mathbf{A} = \mathbf{r}\left[\frac{1}{r}\frac{\partial A_z}{\partial \theta} - \frac{\partial A_\theta}{\partial z}\right] + \boldsymbol{\theta}\left[\frac{\partial A_r}{\partial z} - \frac{\partial A_z}{\partial r}\right]$$

$$+ \frac{1}{r}\mathbf{k}\left[\frac{\partial(rA_\theta)}{\partial r} - \frac{\partial A_r}{\partial \theta}\right]$$

$$\nabla^2 G = \frac{1}{r}\frac{\partial}{\partial r}\left(r\frac{\partial G}{\partial r}\right) + \frac{1}{r^2}\frac{\partial^2 G}{\partial \theta^2} + \frac{\partial^2 G}{\partial z^2}.$$

Spherical Coordinates (r, θ, ϕ)

$$\nabla G = \mathbf{r}\frac{\partial G}{\partial r} + \frac{1}{r}\boldsymbol{\theta}\frac{\partial G}{\partial \theta} + \frac{1}{r\sin\theta}\boldsymbol{\phi}\frac{\partial G}{\partial \phi}$$

$$\nabla \cdot \mathbf{A} = \frac{1}{r^2}\frac{\partial(r^2 A_r)}{\partial r} + \frac{1}{r\sin\theta}\frac{\partial(\sin\theta A_\theta)}{\partial \theta} + \frac{1}{r\sin\theta}\frac{\partial(A_\phi)}{\partial \phi}$$

$$\nabla \times \mathbf{A} = \frac{\mathbf{r}}{r\sin\theta}\left[\frac{\partial(A_\phi \sin\theta)}{\partial \theta} - \frac{\partial(A_\theta)}{\partial \phi}\right]$$

$$+ \boldsymbol{\theta}\left[\frac{1}{r\sin\theta}\frac{\partial(A_r)}{\partial \phi} - \frac{1}{r}\frac{\partial(rA_\phi)}{\partial r}\right]$$

$$+ \boldsymbol{\phi}\left[\frac{1}{r}\frac{\partial(rA_\theta)}{\partial r} - \frac{1}{r}\frac{\partial(A_r)}{\partial \theta}\right]$$

$$\nabla^2 G = \frac{1}{r^2}\frac{\partial}{\partial r}\left(r^2\frac{\partial G}{\partial r}\right) + \frac{1}{r^2\sin\theta}\frac{\partial}{\partial \theta}\left(\sin\theta\frac{\partial G}{\partial \theta}\right)$$

$$+ \frac{1}{r^2\sin^2\theta}\frac{\partial^2 G}{\partial \phi^2}.$$

Maxwell's Wave Equations

$$\nabla \cdot \mathbf{E} = 0$$

$$\nabla \cdot \mathbf{H} = 0$$

$$\nabla \times \mathbf{E} = -\mu\frac{\partial \mathbf{H}}{\partial t}$$

$$\nabla \times \mathbf{H} = \epsilon\frac{\partial \mathbf{E}}{\partial t}$$

$$\frac{\partial^2 \mathbf{E}}{\partial t^2} = c^2\nabla^2\mathbf{E}$$

$$\frac{\partial^2 \mathbf{H}}{\partial t^2} = c^2\nabla^2\mathbf{H}, \text{ where } c = \frac{1}{\sqrt{\epsilon\mu}}.$$

Wallis' Formula

$$\int_0^{\frac{\pi}{2}} \sin^m\theta\cos^n\theta d\theta = \frac{[(m-1)(m-3)\ldots 2 \text{ or } 1][(n-1)(n-3)\ldots 2 \text{ or } 1]}{(m+n)(m+n-2)\ldots 2 \text{ or } 1}\alpha,$$

where,

$$\alpha = \begin{cases} \frac{\pi}{2} & \text{if } m \text{ and } n \text{ are both even} \\ 1 & \text{otherwise.} \end{cases}$$

Index